Symposium of the 71st Tobacco Science Research Conference

Bonita Springs, Florida, USA
28 November – 1 December

Editors:

Ian M. Fearon
Summer N. Hanna
Kathy E. Humphries

ISBN: 978-1-5108-7968-3

Printed from e-media with permission by:

Curran Associates, Inc.
57 Morehouse Lane
Red Hook, NY 12571

Some format issues inherent in the e-media version may also appear in this print version.

Printed by Curran Associates, Inc. (2019)

TSRC Conference Committee
http://www.tsrcinfo.com
info@tsrcinfo.com

Additional copies of this publication are available from:

Curran Associates, Inc.
57 Morehouse Lane
Red Hook, NY 12571 USA
Phone: 845-758-0400
Fax: 845-758-2633
Email: curran@proceedings.com
Web: www.proceedings.com

Recent Advances In Tobacco Science

Volume 43

Engagement Across Disciplines: Tobacco Science in the 21st Century

Symposium Proceedings
71st Meeting
TOBACCO SCIENCE RESEARCH CONFERENCE

November 28 - December 1, 2017
Bonita Springs, Florida USA

Symposium of the
71st Tobacco Science Research Conference

Engagement Across Disciplines:
Tobacco Science in the 21st Century

– Symposium Chair –

Ian M. Fearon

– Editors –

Ian M. Fearon

Summer N. Hanna

Kathy E. Humphries

CONTENTS

Symposium of the
71st Tobacco Science Research Conference

Engagement Across Disciplines:
Tobacco Science in the 21st Century

– CONTRIBUTORS –

Amy E. Brannan
Ian M. Fearon
Derek Mariner
Willie J. McKinney
James Murphy
Cheryl K. Olson
Christopher J. Proctor
Christopher Russell
Donna C. Smith
Christopher Wright

PREFACE

The Program Editorial Committee of the 71st Tobacco Science Research Conference is pleased to present the 43rd volume of the Recent Advances in Tobacco Science publication. Each year the Program Editorial Committee of the Conference selects a theme that highlights a scientific or regulatory issue that is relevant, engaging and thought-provoking for all of the tobacco science community. This year, the committee has chosen "Engagement Across Disciplines: Tobacco Science in the 21st Century" for this year's symposium theme. The symposium discusses the topic of scientific engagement both within and outside of the tobacco industry and looks towards how tobacco industry engagement might take shape in the future. Four distinguished speakers were invited to provide their unique insights into different aspects of tobacco science engagement; within the industry, with the broader scientific community, through the media and with consumers. This publication contains the synopses of the symposium presentations and introductory remarks that include a brief biographical sketch of the symposium speakers. Members of the Program Editorial Committee, Summer Hanna, Kathy Humphries and I, wish to express our sincere appreciation to the speakers, Drs. Chris Proctor, Cheryl Olson, Willie McKinney, and Chris Russell and their colleagues for the significant time and effort spent preparing the publications and presentations.

Ian M. Fearon
Chair, Program Editorial Committee
71st Tobacco Science Research Conference

71st TOBACCO SCIENCE RESEARCH CONFERENCE

November 28 – December 1, 2017
Bonita Springs, Florida USA

Final Program

Tuesday, November 28, 2017

3:00 PM - University of Kentucky CTRP Proficiency Program Workshop
6:30 PM – Welcome Reception hosted by Tobacco Technology, Inc. and its subsidiary eLiquiTech

Wednesday Morning, November 29, 2017

General Session

8:45 **WELCOME:** Tom Cravotta, Tobacco Technology, Inc., 71st TSRC Chair

8:55 **SYMPOSIUM: "Engagement Across Disciplines: Tobacco Science in the 21st Century"**
Chair: Ian M. Fearon, British American Tobacco (Investments) Limited, Southampton, UK **...1**

9:00 **1. INDUSTRY ENGAGEMENT WITH THE SCIENTIFIC COMMUNITY.**
Willie J. MCKINNEY, Amy E. Brannan and Donna C. Smith; Altria Client Services, Richmond, VA, USA **...3**

9:30 **2. THE ROLE OF INDUSTRY IN ENGAGING WITH INDUSTRY TO DEVELOP REGULATORY-APPROVED METHODS FOR TOBACCO AND NICOTINE PRODUCT ASSESSMENT.**
Christopher J. Proctor, Ian M. FEARON, Derek Mariner, Christopher Wright and James Murphy; British American Tobacco (Investments) Limited, Southampton, UK **...9**

10:00 **Break**

10:30 **3. FACTORS AFFECTING MEDIA COVERAGE OF TOBACCO RESEARCH NEWS.**
Cheryl K. Olson; Cheryl K. Olson ScD, LLC, San Carlos, CA, USA **...21**

11:00 **4. VAPERS AND THE VAPING INDUSTRY ARE THE PRIMARY AGENTS OF TOBACCO HARM REDUCTION IN THE UNITED STATES**
Christopher RUSSELL; Centre for Substance Use Research, Glasgow, UK.**....47**

11:30 **Panel discussion with all symposium speakers**

12:00 **Lunch**

1:00 **Poster Session**

5. USE OF FORMALIN IN E-VAPOR PRODUCTS TO MONITOR FORMALDEHYDE DELIVERY. Kathleen SPANGLER, James Wilkinson, Matt Melvin, John H. Miller IV and Georgios D. Karles; Altria Client Services, Richmond, VA, USA...**61**

6. CHARACTERIZATION OF AEROSOLS GENERATED FROM A PROTOTYPE E-VAPOR PRODUCT AND A CAPILLARY AEROSOL GENERATOR. Jingjie ZHANG, Christopher Tucker, K. Monica Lee and Michael Oldham; Altria Client Services, Richmond VA, USA**...62**

7. NEUTRAL RED UPTAKE CYTOTOXICITY ASSAY IN A549 CELLS UNDER DIFFERENT CULTURE CONDITIONS. Bonnie G. COFFA, Utkarsh Doshi, Jingjie Zhang, Willie J. McKinney, K. Monica Lee; Altria Client Services, Richmond, VA, USA and Pooja Desai; Enthalpy Analytical, Richmond, VA, USA**...**62

Note that the 71st TSRC had to be rescheduled due to a hurricane. As a result, several papers were withdrawn, some presentations are not consecutive due rescheduling, and several had to be presented by the author's colleague.

Note that the 71st TSRC had to be rescheduled due to a hurricane. As a result, several papers were withdrawn, some presentations are not consecutive due rescheduling, and several had to be presented by the author's colleague.

Note that the 71st TSRC had to be rescheduled due to a hurricane. As a result, several papers were withdrawn, some presentations are not consecutive due rescheduling, and several had to be presented by the author's colleague.

Note that the 71st TSRC had to be rescheduled due to a hurricane. As a result, several papers were withdrawn, some presentations are not consecutive due rescheduling, and several had to be presented by the author's colleague.

Wednesday Afternoon, November 29, 2017

Session A
NNN
Session Chair: Summer Hanna

2:20 PM
68. THE SCIENTIFIC EVIDENCE FOR THE PROPOSED NNN STANDARD. Kristin MARANO[1], Charlene Liu[1], Geoffrey M. Curtin[1], Tiffany Parms[1], Annette Bachand[2], Sandra I. Sulsky[2], Greg Mariano[2], Tracy Greene[3] and P. Robinan Gentry[3];[1]RAI Services Company, Winston-Salem, NC; USA, [2]Ramboll Environ, Amherst, MA, USA; [3]Ramboll Environ, Monroe, LA, USA...**76**

2:40 PM
69. CRITICAL ASSESSMENT OF THE ANALYTICAL METHODS MEASURING NNN IN SMOKELESS TOBACCO PRODUCTS. Candice K. CUNNINGHAM; RAI Services Company, Winston-Salem, NC, USA ...**77**

3:00 PM
70. GENETIC APPROACHES AND RESOURCES FOR ENABLING COMPLIANCE TO PROPOSED NNN STANDARDS. Ralph E. DEWEY[1], Ramsey S. Lewis[1], Jianli Lu[1], Lucien Bovet[2], Simon Goepfert[2] and Prisca Campanoni[2]; [1]North Carolina State University, Raleigh, NC, USA; [2]Philip Morris International, Neuchatel, Switzerland ...**77**

Session B
Materials/Manufacturing/Finished Product
Session Chair: Thaddeus Jackson

2:20 PM
76. THE POWER OF TACTILITY: TIPPING PAPER IN INTERACTION WITH THE CONSUMER. Michael LINDNER and Eike Schopper; Tannpapier, Traun, Austria ...**79**

2:40 PM
77. AN APPROACH TO IDENTIFY FACTORS RELATING TO CHANGES IN CHEMICAL CONSTITUENTS IN CURED TOBACCO LEAF DURING AGING. Atsushi NAGAI and Daichi Mochizuki; Japan Tobacco, Yokohama, Japan...**80**

3:00 PM
96. AN EVALUATION OF THE VARIABILITY OF HPHCS IN CIGARS AS COMPARED TO CIGARETTES. Karl A. WAGNER, Tammy L. Blake, Matt S. Melvin, Michael J. Morton and Jennifer H. Smith; Altria Client Services, Richmond, VA, USA...**87**

3:20 Break

Session A

3:50 PM
71. NNN LEVELS IN STABLE REDUCED CONVERTER (SRC) AND LC LINES CURED UNDER CONDITIONS THAT FAVOR NNN FORMATION. Marcos LUSSO, Kenworth Lion, Andrew Adams, Whit Morris, Ujwala Warek and James Strickland; Altria Client Services, Richmond, VA, USA ...**78**

4:10 PM
72. N-NITROSONORNICOTINE REDUCTION IN DARK TOBACCO VARIETIES AND SMOKELESS PRODUCT PROTOTYPES. Marcos LUSSO, Andrew Adams, Benjamin Lewis, Timothy Poyner, Kenworth Lion, Lionel DeLoach, Timothy Danielson, James Franke, Ujwala Warek and James Strickland; Altria Client Services, Richmond, VA, USA ...**78**

Session B

3:50 PM
79. PHYSICAL CHARACTERISTICS OF HANDMADE PREMIUM CIGARS: SPECIFICATIONS AND CONSEQUENCES OF SPECIFICATIONS. Beatrice TEILLET, Thomas Verron, Stephane Colard; SEITA, Fleury-les-Aubrais, France...81

4:10 PM
80. WASTE REDUCTION IN THE PRIMARY BY USING MICROWAVE IN-LINE TECHNOLOGY FOR PROCESS CONTROL AND OUTGOING GOODS INSPECTION. Hendrik RICHTER; TEWS Elektronik, Hamburg, Germany ...**81**

Note that the 71st TSRC had to be rescheduled due to a hurricane. As a result, several papers were withdrawn, some presentations are not consecutive due rescheduling, and several had to be presented by the author's colleague.

4:30 PM

73. COMPARISON OF INTRA-CROP YEAR VARIABILITY IN NNN IN TOBACCO AND NNN LEVELS IN SMOKELESS TOBACCO PRODUCTS. Michael J. OLDHAM, David J. Phillips, Jason L. Jordan, Michael J. Morton, Kenworth E. Lion III, Marcos F. Lusso, James E. Franke and James A. Strickland: Altria Client Services, Richmond, VA, USA...**78**

4:50 PM

74. THE ENANTIOMERIC COMPOSITION OF N'-NITROSONORNICOTINE IN AIR-CURED TOBACCO INFLUENCED BY NICOTINE DEMETHYLATION. CAI Bin[1], Huihua Ji[2], Franklin F. Fannin[2], and Lowell P. Bush[2]; [1]Guizhou Academy of Tobacco Science, Guiyang City, China; [2]University of Kentucky, Lexington, KY, USA...**79**

4:30 PM

81. A PRODUCT SURVEY OF 25 PIPE TOBACCOS PURCHASED FROM WEB-BASED MERCHANTS. John H. LAUTERBACH; Lauterbach & Associates, Macon, GA, USA ...**81**

Adjourn

Thursday Morning, November 30, 2017

Session A
eCigarettes/Alternative Cigarettes
Session Chair: Gene Gillman

8:30 AM

83. THERMAL DEGRADATION STUDIES OF ELECTRONIC CIGARETTE LIQUIDS PART 1: A NOVEL ANALYTICAL METHOD TO STUDY A-DICARBONYL FORMATION. Matt S. MELVIN, Karen C. Avery, Regina M. Ballentine, William P. Gardner and Karl A. Wagner; Altria Client Services, Richmond, VA, USA...**82**

8:50 AM

84. THERMAL DEGRADATION STUDIES OF ELECTRONIC CIGARETTE LIQUIDS PART 2: DEVELOPMENT OF A MODEL REACTION SYSTEM USED TO STUDY A-DICARBONYL FORMATION. Matt S. MELVIN, Karen C. Avery, Regina M. Ballentine, William P. Gardner and Karl A. Wagner; Altria Client Services, Richmond, VA, USA...**82**

9:10 AM

85. THE DETERMINATION OF DIACETYL AND ACETYLPROPIONYL IN AEROSOLS FROM ELECTRONIC SMOKING DEVICES USING GAS CHROMATOGRAPHY TRIPLE QUAD MASS SPECTROMETRY. Serban MOLDOVEANU, Amanda G. Hudson and Andrew Harrison; R.J. Reynolds Tobacco, Winston-Salem, NC, USA ...**82**

Session B
Regulation & Quality
Session Chair: Chris Keller

8:30 AM

91. INVESTIGATION OF TSNA FORMATION IN ELECTRONIC CIGARETTE LIQUIDS AND AEROSOL. Xiaohong Cathy JIN, Karen C. Avery, Georgios D. Karles, William P. Gardner, Matt S. Melvin and Karl A. Wagner; Altria Client Services, Richmond, VA USA...**85**

8:50 AM

92. COMPARISON OF RESULTS FROM THREE TECHNIQUES FOR THE ESTIMATION OF E-LIQUID pH-VALUES. John H. LAUTERBACH; Lauterbach & Associates, Macon, GA, USA ...**85**

9:10 AM

93. IMPACT OF CIGAR PHYSICAL VARIABILITY ON CIGAR EXPOSURE USING PROBABILISTIC RISK ASSESSMENT. Felix Ayala-Fierro, Karshak KOSARAJU and Rob Stevens; ITG Brands, Greensboro, NC, USA...**85**

Note that the 71st TSRC had to be rescheduled due to a hurricane. As a result, several papers were withdrawn, some presentations are not consecutive due rescheduling, and several had to be presented by the author's colleague.

9:30 AM
86. ANALYSIS OF (S)- AND (R)-NICOTINE IN VARIOUS NICOTINE SAMPLES AND IN E-LIQUIDS. Serban MOLDOVEANU[1], Gary M. Dull[2], Ross J. Oden[1], Wayne A. Scott[1]; [1]R.J. Reynolds Tobacco, Winston-Salem, NC USA; [2]RAI Innovation, Winston-Salem, NC, USA...**83**

9:50 AM
87. ANALYSIS OF ORGANIC ACID CONTENT IN E-LIQUIDS AND E-CIGARETTE AEROSOL BY HIGH PERFORMANCE LIQUID CHROMATOGRAPHY – ULTRAVIOLET DETECTION (HPLC-UV). Angel RODRIGUEZ-LAFUENTE, Romulo Lodevico, Mehran Sharifi, Peter Joza and William Rickert; Labstat International, Kitchener, ON, Canada **...83**

9:30 AM
94. CONCEPT OF CHARACTERIZATION AND CERTIFICATION OF THE 1R6F REFERENCE CIGARETTE. Socrates Jose P. CAÑETE; University of Kentucky, Lexington, KY, USA **...86**

9:50 AM
95. ALTERNATE MATERIALS AND THEIR POTENTIAL IMPACT ON HPHCS. Michael J. MORTON, Nripa T. Jain, Kathleen H. Fox, Raquel M. Olegario, and Timothy L. Danielson; Altria Client Services, Richmond, VA, USA**...86**

Break

10:40 AM
115. Withdrwan

11:00 AM
89. AN ALTERNATIVE APPROACH TO QUANTITATIVE METHODS: LIMIT TEST FOR ANALYSIS OF DEEMING ANALYTES IN ELECTRONIC CIGARETTES. Salem CHOUCHANE, Darius B. Grissom and Carl J. Adams; Eurofins Lancaster Laboratories, Winston-Salem, NC, USA **...84**

11:20 AM
90. MARKET ENDS E-LIQUIDS AND OPEN DEVICE CHEMICAL EVALUATION. David T. SZABO[1], Amanda G. Hudson[2], and Courtney G. Culbert[1]; [1]RAI Services Company, Winston Salem, NC USA; [2]R. J. Reynolds Tobacco, Winston Salem, NC, USA **...84**

10:40 AM
78. DEVELOPMENT OF A MULTI-STAGE LIQUID EXTRACTION METHOD FOR THE QUANTITATIVE ANALYSIS OF FLAVORS IN TOBACCO FILLER. John A MATHIS, R. A. Stanelle and J.-M. D. Dimandja; U.S. Food and Drug Administration, Atlanta, GA, USA **...80**

11:00 AM
97. EVALUATION OF MAINSTREAM CIGARETTE SMOKE COMPONENTS YIELDS OF DIFFERENT CIGARETTE TYPES BASED ON PRINCIPAL COMPONENT ANALYSIS. LI Zhonghao, Yang Fei, Liu Shanshan, Fan Ziyan, Wang Ying, Bian Zhaoyang, Deng Huimin, Tang Gangling and Hu Quingyuan; China National Tobacco Quality Supervision & Test Center, Zhengzhou, China **...87**

11:20 AM
98. OXIDATIVE STRESS AND OXIDATIVE DAMAGE IN HUMAN LUNG EPITHELIAL CELLS INDUCED BY TOTAL PARTICULATE MATTER EXTRACTIONS FROM THREE DIFFERENT TOBACCOS. KANG Yu, Qiao Liangiun, Hu Qiao, Hua, Chenfeng, Xie Fuwei, Shang Pingping, Li Xiang and Zhao Junwei; Zhengzhou Tobacco Research Institute, Zhengzhou, China **...87**

11:40 AM
99. SIMULTANEOUS DETERMINATION OF 22 AMADORI COMPOUNDS IN TOBACCO LEAVES BY LC-MS/MS. WANG Xiaoyu, Liu Leiyu, Wang Sheng, Qin Yaqiong, Jia Yunzhen, Cui Huapeng, Pan Lining, Xie Fuwei and Liu Huimin; ; Zhengzhou Tobacco Research Institute, Zhengzhou, China **...88**

Lunch

Note that the 71st TSRC had to be rescheduled due to a hurricane. As a result, several papers were withdrawn, some presentations are not consecutive due rescheduling, and several had to be presented by the author's colleague.

1:00 PM – Poster Session – See Poster Numbers 5-67 & 105 on Wednesday afternoon

Thursday Afternoon, November 30, 2017

Session A
eCigarettes/Alternative Cigarettes/Human Smoking
Session Chair: Candice Cunningham

2:00 PM
100. ELECTRONIC CIGARETTE AEROSOL DYNAMICS IN A PHYSICAL MODEL OF THE ADULT HUMAN ORAL/PHARYNGEAL CAVITY. Weiling LI[1], Qiang Wang[2]; Nicolas Castro[1], Jingjie Zhang[1], Yezdi Pithawalla[1], Michael Oldham[1], Peter Lipowicz[1] and Ali Rostami[1]; [1]Altria Client Services Richmond, VA, USA; [2]Lancaster Laboratories, c/o Altria Client Services, Richmond, VA, USA ...**88**

2:20 PM
127. FUMEX: LIGHT SCATTERING SENSOR FOR THE ANALYSIS OF ELECTRONIC CIGARETTE AEROSOLS. Weiling LI[1], Qiang Wang[2]; Peter Lipowicz[1], Wilhelm. Dunkhorst[3] and Wolfgang Koch[3]; [1]Altria Client Services, Richmond, VA USA, [2]Lancaster Laboratories, c/o Altria Client Services, Richmond, VA USA, Fraunhofer ITEM, Hannover, Germany...**96**

2:40 PM
103. THE MEASUREMENT OF ADDICTIVENESS AND ATTRACTIVENESS. Neil SHERWOOD; Neil Sherwood Consulting, Commugny, Switzerland ...**89**

3:00 PM
104. ADAPTING TRADITIONAL HUMAN ABUSE LIABILITY TESTING TO TOBACCO PRODUCTS. Graham K. Wood, Graham K. Wood and Michael C. KONG; Altasciences Clinial Research, Laval, QC, Canada ...**89**

3:50 PM
88. NICOTINE PHARMACOKINETICS OF ELECTRONIC CIGARETTES: EXPERIMENTAL DATA AND A REVIEW OF THE LITERATURE. Ian M. FEARON[1], Alison Eldridge[1], Nathan Gale[1], Mike McEwan[1], Paul Nelson[2], Elaine Round[2] and Mitch Stiles[2]; [1]British American Tobacco (Investments) Limited, Southampton, UK; [2]RAI Services Company, Winston-Salem, NC USA ...**84**

Session B
Regulatory Research to Support Tobacco Product Submissions
Session Chair: Geoff Curtin

2:00 PM
108. COMPREHENSION AND PERCEPTIONS OF REDUCED RISK INFORMATION FOR CAMEL SNUS. Kimberly FROST-PINEDA[1], Janine Pillitteri[2], Saul Shiffman[3], and Geoffrey Curtin[1]; RAI Services Company, Winston-Salem, NC, USA; [2]Pinney Associates, Burlington, VT, USA; [3]Pinney Associates, Pittsburgh, PA, USA ...**91**

2:20 PM
109. REDUCED RISK INFORMATION FOR CAMEL SNUS: PROJECTING LIKELIHOODS OF USE AMONG CURRENT SMOKERS, FORMER SMOKERS AND NEVER TOBACCO USERS. Geoffrey CURTIN[1], Karen Gerlach[2] and Saul Shiffman[2], [1]RAI Services Company, Winston-Salem, NC, USA; [2]Pinney Associates, Pittsburgh, PA, USA. ...**91**

2:40 PM
110. MODELING THE POPULATION HEALTH EFFECTS OF CAMEL SNUS WITH REDUCED RISK INFORMATION. Annette Bachand[1], Sandra Sulsky[1], Saul Shiffman[2] and Geoffrey CURTIN[3]; [1]Ramboll Environ U.S., Amherst, MA, USA; [2]Pinney Associates, Pittsburgh, PA, USA; RAI Services Company, Winston-Salem, NC, USA ...**92**

3:00 PM
111. FLAVORED E-CIGARETTE USE AMONG U.S. ADULTS: RESULTS FROM TWO NATIONAL SURVEYS. Mimi KIM[1], Mark Sembower[2], Saul Shiffman[2]; and Geoffrey Curtin[1]; RAI Services Company, Winston-Salem, NC, USA; [2]Pinney Associates, Pittsburgh, PA, USA ...**92**

Break

3:50 PM
112. VARIATIONS IN INTENSITY OF E-CIGARETTE USE, SMOKING HISTORY, AND DEMOGRAPHICS AMONG PAST-30-DAY E-CIGARETTE USERS. Saul Shiffman[1], Mark Sembower[1], Mimi KIM[2] and Geoffrey Curtin[2]; [1]Pinney Associates, Pittsburgh, PA, USA; [2]RAI Services Company, Winston-Salem, NC, USA....**93**

Note that the 71st TSRC had to be rescheduled due to a hurricane. As a result, several papers were withdrawn, some presentations are not consecutive due rescheduling, and several had to be presented by the author's colleague.

4:10 PM

106. A STUDY TO EXAMINE CHANGES IN EXPOSURE TO CIGARETTE SMOKE CHEMICALS WHEN A SMOKER SWITCHES TO USING A TOBACCO HEATING PRODUCT. PART I: STUDY DESIGN. Nathan GALE[1], Mike McEwan[1], Alison Eldridge[1], Oscar M. Camacho[1], John McAughey[1], James Murphy[1], Chuan Liu[1], Christopher J. Proctor[1], Ian M. Fearon[1], Neil Sherwood[2], Edward Bowen[3], Simon McDermott[3] and Emma Holmes[3]; [1]British American Tobacco (Investments) Ltd, Southampton, UK; [2]Neil Sherwood Consulting, Commugny, Switzerland; [3]Covance, Leeds, UK...**90**

4:30 PM

107. A STUDY TO EXAMINE CHANGES IN EXPOSURE TO CIGARETTE SMOKE CHEMICALS WHEN A SMOKER SWITCHES TO USING A TOBACCO HEATING PRODUCT. PART II: STUDY FINDINGS. Nathan GALE[1], Alison Eldridge[1], Mike McEwan[1], Graham Errington[1], James Murphy[1], Christopher J. Proctor[1], Ian M. Fearon[1], Simon McDermott[2], James Glew[2] and Andrew Hedge[2]; [1]British American Tobacco (Investments) Ltd, Southampton, UK; [2]Covance, Leeds, UK ...**90**

4:10 PM

113. COMPARISION OF BACKGROUND RISKS WITH RISKS ESTIMATED FROM CONSTITUENTS IN TOBACCO PRODUCTS. Charlene LIU and Kristin Marano, RAI Services Company, Winston-Salem, NC, USA ...**93**

Adjourn

5:00 PM – TSRC Business Meeting – All attendees are encouraged to attend.

6:30 PM – Pre-banquet Reception

Note that the 71st TSRC had to be rescheduled due to a hurricane. As a result, several papers were withdrawn, some presentations are not consecutive due rescheduling, and several had to be presented by the author's colleague.

Friday Morning, December 1, 2017

Combined Session
Agronomy
Method Development & Smoke Chemistry

Session Chairs: Kathy Humphries & Ian Fearon

8:30 AM
116. QUANTIFYING TOBACCO LEAF COMPOUNDS BY VISIBLE LIGHT ANALYSIS. Makoto YOSHIZAKI; Japan Tobacco, Yokohama, Japan **...93**

8:50 AM
117. INVESTIGATION ON THE ROLE OF COPPER (II) ON NITROSAMINE FORMATION FROM NICOTINE. Syed HAQUE and Socrates Jose P. Canete; University of Kentucky, Lexington, KY, USA **...94**

9:10 AM
118. PON AND NNK CHANGES IN CURED BURLEY TOBACCO WITH STORAGE. Ying WU, Huihua Ji, Franklin Fannin and Lowell Bush; University of Kentucky, Lexington, Kentucky, USA **...94**

9:30 AM
123. THE REGULATION OF MICROORGANISM ON THE FORMATION OF TSNAS IN TOBACCO. ZOU Congming, Xia Zhenyuan, Lei Liping, Huang Wei, Zhao Gaokun., Jin Yan and Wu Qun; Yunnan Academy of Tobacco Agricultural Sciences, Kunming, China **...94**

Break

10:20 AM
124. ARE AVAILABLE TEST METHODS FOR THE DETERMINATION OF AMMONIA IN MAINSTREAM CIGARETTE SMOKE FIT FOR THE ANALYSIS OF CIGARS? Yevgeniya PREPELITSKAYA, Kathleen Sprangler, Jennifer H. Smith, Karen Avery, James Wilkinson, Matt Melvin and John H. Miller IV; Altria Client Services, Richmond, VA USA **...95**

10:40 AM
125. DETERMINATION OF HARMFUL AND POTENTIALLY HARMFUL CONTAMINANTS IN TOBACCO SMOKE USING STANDARD REFERENCE MATERIAL 3222. Walter B. WILSON and Lane C. Sander, National Institute of Standards and Technology, Gaithersburg, MD, USA**...95**

11:00 AM
129. A COMPUTATIONAL MODEL TO CHARACTERIZE THE VITROCELL® CELL EXPOSURE SYSTEM FOR EVALUATION OF AEROSOLS. Nicolas CASTRO[1], Ali A. Rostami[1], Michael J. Oldham[1], Yezdi B. Pithawalla[1]; Francesco Lucci[2] and Arkadiusz Kuczaj[2]; [1]Altria Client Services, Richmond, VA, USA; [2]Philip Morris International, Neuchâtel, Switzerland **...96**

Adjourn

Note that the 71st TSRC had to be rescheduled due to a hurricane. As a result, several papers were withdrawn, some presentations are not consecutive due rescheduling, and several had to be presented by the author's colleague.

INTRODUCTION TO THE SYMPOSIUM

Ian M. Fearon

British American Tobacco
Southampton, United Kingdom

Scientific engagement is a key part of all of our working lives as tobacco science researchers. Traditionally, the main route of our engagement and scientific communication has been through peer-reviewed publications and both oral and poster presentations at scientific conferences such as TSRC. Recent years however have seen a shift in this pattern, and our science is being more broadly communicated and through new channels such as the media and through direct communications with the consumers of the products we develop and make. Increasingly also, we are required to engage and communicate with regulatory authorities who oversee the development, production and marketing of nicotine and tobacco products.

The choice of symposium topic, 'Engagement Across Disciplines: Tobacco Science in the 21st Century', was chosen by the Program Editorial Committee to highlight the traditional and new routes of engagement and scientific communication. We hope that this will be of interest to all attendees at the TSRC conference across the range of scientific interests of attendees, from plant biology, through product development innovation, and on to consumer engagement and post-market research. Of importance, the Symposium will also pay attention to industry collaboration, which many see as a critical element as the industry may be better placed if it acts in concert in many disciplines of scientific research and communication.

Dr. Willie McKinney, Vice President of Regulatory Sciences at Altria Client Services, will open the Symposium with a talk entitled 'Industry Engagement with the Scientific Community'. Dr McKinney has had a distinguished career as both an academic and an industry scientist, and is currently serving on FDA's Tobacco Product Scientific Advisory Committee (TPSAC) as the non-voting representative of the tobacco manufacturing industry. His experience will enable him to authoritatively discuss how the industry can benefit from its interactions with stakeholders in the broader scientific community, and his presentation will provide a unique insight into the drivers of academic science and university scientists and discuss how a better relationship can help us work towards a common goal of tobacco harm reduction.

Dr. Chris Proctor, Chief Scientific Officer at British American Tobacco, will follow with a paper entitled 'The role of industry consortia to develop regulatory-approved methods for tobacco and nicotine product assessment'. Dr. Proctor was more than 25 years of experience of the tobacco industry and regulatory engagement, and in recent years has been a key driver in cross-industry collaboration in areas such as in vitro models of toxicology and population modelling. His presentation will discuss why such collaborative efforts are important in a modern tobacco and nicotine industry and uses case studies from other industries to highlight the benefits that the regulated industry can gain from working together on non-competitive areas of scientific research.

Following on from this, Dr. Cheryl Olson will talk about 'Factors affecting media coverage of tobacco research news'. Dr. Olson is an independent consultant and an internationally-renowned expert on using media to change behavior, particularly in the area of promoting mental and physical health. Drawing on her experience of working across a range of different areas and with diverse companies such as VTech, Philip Morris USA and Activision, she will provide a unique insight into the quality, nature and motivations of press coverage of scientific findings, using recent examples of the reporting of studies on the health effects of tobacco, nicotine or e-cigarettes to reveal emerging patterns and trends. Her presentation will conclude by touching on how researchers can act to improve the likelihood and quality of media coverage of research findings.

The final presentation of the Symposium will be made by Dr. Christopher Russell, an independent behavioural psychology Senior Research Fellow at the Centre for Substance Use Research in Glasgow, Scotland. Dr Russell leads the Centre's tobacco harm reduction research studies, with a particular focus on how vapers and e-cigarettes are helping smokers to quit smoking. In recent years, Dr. Russell has been a key figure in understanding the harm reduction potential and risks/benefits of e-cigarette use and in particular he has used several multi-national surveys to understand the drivers, motivations and barriers to smokers' using electronic cigarettes. With this experience his talk entitled 'Vapers and the vaping industry are the primary agents of tobacco harm reduction in the United States' will give his insight into how the manufacturers and consumers of e-cigarettes are driving tobacco harm reduction efforts and how current smokers, who are potential consumers of harm reduction products, can be assisted in quitting smoking using online electronic resources and peer-to-peer support.

Following the presentations, the symposium will conclude with an open panel discussion. All of the speakers will be invited back to the stage where they will field questions pertaining to scientific engagement and communication. We encourage attendees to make the best use of the speakers' collective wealth of expertise and knowledge in this area and to actively participate in this question and answer session.

The Editorial Committee would like to recognize the speakers and organisations for their contributions to the symposium and extend our gratitude for their participation. We hope that the symposium is found to be informative, insightful and inspiring to conference participants and that the topics presented in the symposium will promote dialogue and collaboration towards our common interest of communicating and engaging with different stakeholders on our scientific research.

Ian M. Fearon
71st TSRC Symposium Chair

INDUSTRY ENGAGEMENT WITH THE SCIENTIFIC COMMUNITY

Willie J. McKinney, Amy E. Brannan and Donna C. Smith
Altria Client Services LLC
Richmond, Virginia USA

Abstract
Collaborations between industry scientists and the broader scientific community are common. Tobacco industry collaboration with public health scientists presents unique opportunities and challenges. The convergence of i) an FDA regulatory environment where FDA can make science and evidence based decisions about tobacco products and communications; ii) innovations by tobacco companies and concurrent evolving adult tobacco consumer expectations; and iii) divergent perspectives in the public health community relative to tobacco harm reduction all serve to create a nexus for tobacco industry, FDA and academic scientists to effectively collaborate. We all have a role advancing harm reduction. For example, industry and academic scientists may gather the scientific evidence to demonstrate the risk reduction potential of new innovative tobacco products. Their knowledge of cigarette smoke attributable-risk and adult tobacco consumer preferences is important for this analysis. Public health, including FDA, play a pivotal role in advancing harm reduction by educating adult smokers about the identified risk differential between tobacco products so that adult tobacco consumers can make informed decisions. Although productive, these collaborations face several challenges. Most notably, the history of the tobacco industry may sometimes negatively impact effective scientific engagement efforts. However, regardless of potential barriers, several new forums are facilitating productive communications between academic harm reduction enthusiasts, pessimists, public health and the tobacco industry. These forums serve to diminish barriers such as a lack of trust. This paper will present some of the conditions necessary for successful and productive tobacco industry engagement with the scientific community.

Introduction
A renewed focus on reducing the harm caused by cigarettes is providing a unique platform for new collaboration between a tobacco product manufacturer and the scientific community (academic bench and public health scientists). These collaborations have the potential to positively impact public health. With the ever-increasing importance of innovation as an instrument of cigarette harm reduction, it is imperative that a tobacco industry member and the scientific community find common ground. The convergence of i) an FDA regulatory environment where FDA can make science and evidence based decisions about tobacco products and communications; ii) innovations by tobacco companies and concurrent evolving adult tobacco consumer interest and expectations; and iii) divergent perspectives

in the public health community relative to tobacco harm reduction all serve to create a nexus for scientists in the tobacco industry to collaborate with the scientific community.

Background

While significant value may be gained from such collaborations, there are significant barriers to most industry/academic collaboration. The organizational drivers of success that exist in industry differ from academia. In addition, there are clearly different institutional norms governing public and private knowledge management.

For example, unlike industry, the process of creating public knowledge has been central to the growth of universities. In 1973, Merton established four sets of institutional imperatives that comprise the ethos of modern science. These imperatives are now known as Mertonian norms and researchers at academic institutions have been firmly rooted in these norms for centuries. These norms are the framework by which academics perceive the work that they do. These norms can be summarized as:

1. Universalism – scientific validity is independent of from where the science comes
2. Communalism – science should be open and free to promote collaboration
3. Disinterestedness – institutions of science do work for the advancement of science, not the advancement of individuals
4. Organized skepticism – scientific claims should be scrutinized both in the methods used and by the institutional codes of conduct of the institution that made the scientific claim

These norms are organizational in nature and describe attitudes and behavior of academia.

Although Mertonian norms are foundational for academic institutions, economic and social forces outside the science system itself may play a powerful role in shaping scientists and science (Freeman, 1999). For example, government funded research is applied, practically oriented, and focused on solving general social, technical or economic problems (Pavitt, 2001).

In contrast to the open nature of the academic science system described above, the primary motivation of industry knowledge creation activities is to develop products and intellectual property to gain advantage over competitors (Chesbrough, 2006).

The differences in organizational drivers, however, do not mean that there is not common ground between the two institutions. Table 1 on the next page (reproduced from Ivascu *et al.* 2016) shows that in successful collaborations between industry and academia, a real potential exists to meet the organizational needs of both partners while producing significant value for society.

Table 1. Comparing organization drivers between academia and industry

Academic Goals	Shared Goals	Industry Goals
Public mission	Creating value for society	Shareholder value
Publications	Reputation	Revenue
Project research	Research	Practical research
Theoretical drivers	Science driven	Results driven
Shared resources	Competitiveness	Private resources
Sharing results	Value	Retain results
Creating knowledge	Sharing knowledge	Capturing knowledge
Open source	Collaborative innovation	Private source
Investigator needs	Consumer needs	Market needs
Education	Exchange know-how	Retain know-how

If collaborations between academia and industry in general are challenging, collaborations between academia and a tobacco industry member are especially tenuous due to the history of conflict and mistrust. For example, in an article from Public Health Reports in 2005, Lisa Bero outlines six strategies that are what she refers to as the manipulation of research by the tobacco industry:

1. Fund research that supports the interest group position.
2. Publish research that supports the interest group position.
3. Suppress research that does not support the interest group position.
4. Criticize research that does not support the interest group position.
5. Disseminate interest group data or interpretation of risk in the lay press.
6. Disseminate interest group data or interpretation of risk directly to policy makers.

Some members of the scientific community–including academic bench and public health scientists–still do not trust the integrity of scientific data generated by members of the tobacco industry. Lack of trust and other barriers such as organizational policies that require institutions to have bans on receiving money from tobacco companies in order to receive their funding and refusal of certain journals to publish science generated or funded by tobacco industry members have made collaborations more difficult, if not impossible.

Considerations

In order to achieve the common goal of tobacco harm reduction, we must acknowledge that the scientific questions and challenges associated with tobacco harm reduction (*e.g.*, continuum of risk) are extremely diverse, complex and require the collective knowledge of the tobacco industry and scientific community.

In addition, it appears that trust may very well be the most important factor and the biggest barrier to fruitful collaborations between industry and academia. These collaborations can be well formed by professional and personal relationships (Bruneel *et al.*, 2010). This suggests that a tobacco product manufacturer and the scientific community should begin the journey of building trust by understanding the different goals and incentives of one another and focus on face-to-face interactions. Specifically, scientists in the tobacco industry and academia must leverage opportunities to engage by attending conferences, presenting science and inviting questions and critique. Through these exchanges we begin to develop professional relationships, respect, healthy skepticism and, hopefully, eventually trust.

Each organization can build trust by making long-term investments in collaborations that are forged one small project at a time focused on common goals such as "creating value for society." Scientists in the tobacco industry and academia may generate scientific evidence to demonstrate the risk reduction potential of new innovative tobacco products. Their shared knowledge of cigarette smoke attributable-risk and adult tobacco consumer preferences is important for this analysis. Subsequently, government institutions like FDA may utilize peer-reviewed science and evidence to make decisions about tobacco product risk and advance harm reduction by educating adult smokers about the identified risk differential between tobacco products so that adult tobacco consumers can make informed decisions.

We must address the barriers that impede the common goal of sharing scientific information via publications. Policies that impede the ability of either organization to publish do not benefit public health–in fact, these exclusionary policies may well serve to delay significant advances in harm reduction. Through publication, the tobacco industry and scientific community demonstrate their scientific capabilities as well as their dedication to openness and transparency.

Conclusion

Is the common goal of tobacco harm reduction sufficient to overcome a history of mistrust of research funded or conducted by tobacco product manufacturers? In our view, collaboration between a member of the tobacco industry and the scientific community seems possible; especially given the current regulatory environment, adult tobacco consumer acceptance of innovative tobacco products and a common goal of tobacco harm reduction.

References

1. Bero L (2005). Tobacco Industry Manipulation of Research. Public Health Reports. 120(200-208).
2. Bruneel J, D'Este P, Salter A. (2010). Investigating the factors that diminish the barriers to university-industry collaboration. Research Policy 39(858-868).
3. Chesbrough H (2006). Open Business Models: How to Thrive in the New Innovation. Harvard Business School Press. Boston, Massachusetts.
4. Freeman C (1999). The social function of science. In: Swann B, Aprajamian F (Eds), Bernal JD: A Life in Science and Politics. Verson, London, New York.
5. Ivascu L, Cijaliu B, Draghici A (2016). Business model for the university-industry collaboration in open innovation. Procedie Economics and Finance 39(674-678).
6. Pavitt K (2001). Public policies to support basic research: What can the rest of the world learn from US theory and practice? (And What They Should Not Learn). Industrial and Corporate Change 10(761-779).

THE ROLE OF INDUSTRY CONSORTIA TO DEVELOP REGULATORY-APPROVED METHODS FOR TOBACCO AND NICOTINE PRODUCT ASSESSMENT

Christopher J. Proctor, Ian M. Fearon, Derek Mariner,
Christopher Wright and James Murphy
British American Tobacco
Southampton UK

Abstract

Consensus methods for product assessment are commonplace across many industry sectors and often form the basis of regulatory approaches. In the pharmaceutical industry for example, the foundations for the assessment of the safety, quality and efficacy of new medicines were laid almost 30 years ago in cross-industry collaboration with regulators, leading to the inception of the ICH guidelines which are now applied globally.

Recently, in the tobacco industry, when an EU expert scientific panel advised that more scientific data was needed on specific additives used in cigarettes, it suggested that tobacco manufacturers could carry out joint studies to collect the required data. This is a similar approach to that taken with chemical manufacturers to achieve compliance with European chemical regulations.

There are some long-standing cross-industry fora that work on consensus methods, such as the Cooperation Centre for Scientific Research Relative to Tobacco (CORESTA). More recently the Institute for *In Vitro* Sciences (IIVS) hosted a workshop attended by industry scientists and FDA to establish principles for conducting *in vitro* pre-clinical assessment techniques. In Europe a broad stakeholder group comprising manufacturers, regulators and consumer groups was involved in developing the British Standards Institute (BSI) Publicly Available Specification for e-cigarettes and the Association Française de Normalisation (AFNOR) standards. Yet in other areas, such as the development of dynamic population models, individual companies have developed their own models independently of one another.

This paper looks at the importance of cross-industry collaboration in non-competitive contexts, the value of consensus standards for both industry and regulators, and some of the conditions necessary to develop standards trusted by all interested parties.

Introduction

Those outside the tobacco and nicotine industry may find it somewhat surprising that relatively little pre-market collaboration takes place, particularly in the field of developing product assessment tests for regulatory purposes. From the dual perspectives of regulators finding it easier to interpret data when that data is produced through standardised methodologies to the cost-effectiveness for industry of combining resources in developing and validating methodologies, and thus improving the likelihood of gaining regulatory acceptance, it would seem sensible that industries should collaborate in the pre-competitive space.

One of the key underlying reasons for this may be the historic lack of trust in the science undertaken by the tobacco industry. Internationally, when the World Health Organization (WHO) was considering as part of the Framework Convention on Tobacco Control (FCTC) which constituents in cigarette smoke might be included for testing and how such testing should be performed, it took the view that methods previously developed by the industry through CORESTA and the International Organization for Standardization (ISO) Technical Committee 126 were not to be trusted. Consequently, the FCTC Conference of the Parties requested the WHO Tobacco Free Initiative to set up and fund an international consortium of laboratories (Tobacco Laboratory Network, TLN) to develop WHO-approved methods for a limited set of constituents (1). This effort took several years and the methods recommended by WHO are, in most cases, similar to those previously recommended by CORESTA and which have formed the basis for ISO standards.

There are parallels with car emissions testing, in which methods developed in the 1960s and 1970s with the involvement of environmental agencies and car manufacturers are considered today as not being fit-for-purpose or open to misuse (2). Despite early involvement of regulators with testing regimes, there is a public perception that commercial interests are not necessarily in the best interests of public safety.

As a result of changing regulatory requirements, the number and type of scientific studies important to the tobacco and nicotine industry has changed substantially over the past few years. Historically the predominant focus was on analytical chemical methods to assess cigarette emissions. However, with the introduction of the Tobacco Control Act in the United States (3) and its public health standard concerning both individual risk and population risk, a set of new scientific studies are required. While commonplace in some other sectors, studies on abuse liability, risk perception, and comprehension are required prior to launch for some of the regulated pathways to obtaining a marketing order through the FDA. Relatively uncommon to other sectors is the need to develop population models that predict the long-term impact of the introduction of a new product on both product users and non-users.

There are, as yet, no standardised and regulatory accepted methods for testing the addictiveness, attractiveness, or the toxicity of tobacco and nicotine products (4,5). Even in pharmacokinetic studies of nicotine uptake, results can vary dramatically depending on the protocol and the volunteer participants (6,7). While there are adaptations of standardised regulatory tests such as the Ames test, these give limited insights into toxicity. 21st Century toxicological approaches that are being adopted by some regulators (8) have the advantage of giving better insights and substantially reducing the historic reliance on animal tests. The use of systems toxicology and systems biology may become more important, as will the development of biomarkers of biological effect that could predict disease risk in shorter periods than the time it takes to develop chronic diseases.

So, there is no better time for the industry to consider ways in which it might collaborate within itself and with other parties on developing standardised testing regimes for a wide range of end-points.

Industry collaboration in other sectors

(a) Collaboration through Standards Organisations

In virtually all sectors the concept of developing consensus standards with industry and others is accepted. Indeed, this is the way of working for ISO. ISO lists a large number of "Organizations in co-operation with ISO," including as examples many industry collaborative bodies such as the Toy Industries of Europe, the European Federation for Precast Concrete, the Personal Care Association and the European Federation of Pharmaceutical Industries and Associates alongside civil society groups (such as the Framework Convention Alliance in the tobacco sector), learned societies, and other expert groups.

As ISO notes "It started with the obvious things like weights and measures, and over the last 50 years has developed into a family of standards that cover everything from the shoes we stand in, to the Wi-Fi networks that connect us invisibly to each other." (9)

Addressing all these and more, International Standards mean that "consumers can have confidence that their products are safe, reliable and of good quality. ISO's standards on road safety, toy safety and secure medical packaging are just a few of those that help make the world a safer place." (9)

ISO state that regulators and governments count on ISO standards to help develop better regulation, knowing they have a sound basis thanks to the involvement of globally-established experts." (9)

ISO develops standards through Technical Committees (TCs) and recommends broad stakeholder participation where available. For example, ISO TC 34, working on Food Products, has 74 participating members, 64 observing members and 23 international organisations with Liaison status, and has developed around 840 published ISO standards. The members of TCs are national standards organisations from around the world. For example, The American National Standards Institute (ANSI) describes itself as being; "comprised of a broad range of businesses and industrial organizations, standards setting and conformity assessment bodies, trade associations, labor unions, professional societies, consumer groups, academia, and government organizations for the purpose of enhancing global business competitiveness and improving the quality of life for the world's citizens." (1)0

It is commonplace for national standards organisations to invite industry experts to help in the development and validation of new standards.

(b) International harmonisation

In the pharmaceutical sector there are many good examples of industry non-competitive collaboration. The International Council for Harmonisation of Technical Requirements for Pharmaceuticals for Human Use (ICH) brought together the regulatory authorities and pharmaceutical industry to discuss scientific and technical aspects of drug registration. Since its beginnings in 1990, ICH has gradually evolved in response to an increasingly global footprint of drug development. ICH's stated mission is to achieve greater harmonisation worldwide to ensure that safe, effective, and high quality medicines are developed and registered in the most resource-efficient manner (11).

Harmonisation of regulatory requirements was pioneered in Europe in the 1980s, as the European Community moved towards the development of a single market for pharmaceuticals. The success achieved in Europe showed that harmonisation was feasible. While discussions were held between Europe, Japan and the US on possibilities for harmonisation, it was a meeting of the WHO Conference of Drug Regulatory Authorities (ICDRA), in Paris, in 1989, that brought plans to life. This resulted in the authorities approaching the International Federation of Pharmaceutical Manufacturers and Associations (IFPMA) to discuss a joint regulatory-industry initiative on international harmonisation (11).

The Terms of Reference resulted in separating the efforts for harmonisation between Safety, Quality and Efficacy to reflect the three criteria which are the basis for approving and authorising new medicinal products (11).

The initiative continues to progress, and agreement was gained that clinical trials run under Good Clinical Practice conducted in one ICH region could be used in other ICH regions in 2016, and GCP (R2) reflected on the emerging increase in multi-centre studies and the need for adequate oversight (11,12).

After a series of international food safety issues in the 1990s (for example, BSE, Dioxins, Listeria) and to address a lack of consistency of standards and food industry safety assurance procedures, the Global Food Safety Initiative (GFSI) was established by the CEOs of the world's food retailers (13). They worked together through their independent international network CIES - The Food Business Forum, now the Consumer Goods Forum (CGF). The GFSI has 4 objectives:

"1. Reduce food safety risks by delivering equivalence and convergence between effective food safety management systems.
2. Manage cost in the global food system by eliminating redundancy and improving operational efficiency.
3. Develop competencies and capacity building in food safety to create consistent and effective global food systems.
4. Provide a unique international stakeholder platform for collaboration, knowledge exchange and networking." (13)

The approach taken was to establish benchmarking for food safety certification programmes such as the UK British Retail Consortium Food Safety Standard that was first introduced in 1998, the International Food Standard (IFS) developed in Europe and the Safe Quality Food (SQF) Standard developed in North America (14,15).

GFSI has developed to become more than just a benchmarking organisation. Its collaborative approach sees international food safety experts from the entire supply chain meet at Technical Working Group and Stakeholder meetings, conferences, and regional events. The key purpose is stated as sharing knowledge and promoting a harmonised approach with a shared vision of safe food for consumers everywhere. (13)

(c) Collaborations on specific health issues

Some consortia have been set up to tackle specific disease issues. One example is not-for-profit charities such as the Dementia Consortium (16). The Consortium is a cross-sector drug discovery collaboration between Alzheimer's Research UK, MRC Technology and the pharmaceutical companies Eisai, Lilly and Astex. The major companies are said to contribute knowledge, resources, and capital investment to help fund projects. The consortium states that by bringing expertise and providing focused investment, the gap between academic research and the pharmaceutical industry can be bridged in the search for drugs that slow the development of neurodegenerative diseases that cause dementia.

Tackling the issue of tropical disease, a consortium was established, WIPO Re:Search (17), to accelerate discovery and development of medicines, vaccines and diagnostics for neglected tropical diseases. Sponsored by the World Intellectual Property Organization (WIPO), the consortium includes pharmaceutical companies, leading research institutions and nongovernmental organizations. In this example, the member organisations provide intellectual property and expertise on a royalty-free basis to qualified researchers across the global health research community who focus on neglected tropical diseases. Products developed using WIPO Re:Search data are given royalty-free licenses when distributed in least developed countries.

(d) Collaborations with learned societies

Learned societies can also play a role in bringing common ground to marketplace competitors. For example, the American Chemical Society's Green Chemistry Institute Pharmaceutical Roundtable sees member companies who are part of the ACS GCI Pharmaceutical Roundtable come together to look at new ways to improve process efficiency and product quality through green chemistry and engineering. By working together, the Roundtable is said to provide leadership and influence throughout the industry and supply chain (18).

More broadly focused, the ACS Green Chemistry Innovations forum seeks to connect the Green Chemistry community, from businesses seeking innovative solutions to inventors seeking to find applications for their technologies (18).

(e) Collaborations on new technologies

Consortia have been formed between companies in an industrial sector as a response to a common need. For example, the Open Automotive Alliance (19) is a consortium of automotive manufacturers (such as Audi, Fiat and Ford) and technology companies (such as LG and Google) with a goal of bringing the Android platform into cars in a safe and efficient way.

In another automobile example, German car manufacturers joined to form a company, "Here", which captures location content such as road networks, buildings, parks and traffic patterns. It then sells or licenses that mapping content, along with navigation services and location solutions to other businesses such as Alpine, Garmin, BMW, Oracle and Amazon. (20)

In pharmaceuticals, ICH was deemed successful and has encouraged the sector to take a variety of additional initiatives. For example, the International Consortium for Innovation and Quality in the Pharmaceutical Sector (IQ) is a not-for-profit organization of pharmaceutical and biotechnology companies with the mission of advancing science and technology to improve the capability of member companies with the stated aim of developing transformational solutions that benefit patients, regulators and the broader R&D community. Members include major pharmaceuticals such as Pfizer and Astra-Zeneca and smaller focused companies such as Seattle Genetics, specialists in antibody drug conjugate cancer therapies (21).

The Organization for Economic Collaboration and Development (OECD) is the host sponsor for an international effort to develop Adverse Outcomes Pathways (AOPs) (22). Organisations, including industry, agree to host and develop an AOP for a specific disease outcome, and then through the contributions of anyone connected into the system and under the governance of the OECD, the AOPs are improved and developed further. Once developed the key events on the pathways can be evaluated through specific *in vitro* models, which in turn can evolve into methods for evaluating the potential risks of new products. In a similar approach, IBM's Improver, supported by Philip Morris International, uses internet crowd-sourcing to improve and validate the networks in systems biology models (23).

Industry collaboration from the perspective of the regulator

As can be seen in the above, many of the consortia developed by various sectors have a contribution from regulators, and are responsive to either societal or regulatory needs. From a regulator's perspective, there are many advantages in having the sector it regulates work in consort on issues it would like to have addressed. Consortia help with efficiency of resources and speed and also allow regulators to consider one solution rather than having to decide between several presented by different members of a particular industry.

Elements of successful collaborations

Although the above gives just some examples of successful collaborations involving industrial sectors, some general themes emerge. The collaborations arise from a clear purpose, whether that be the acceleration of Green Chemistry or the introduction of international standards for food safety. They often involve regulators or government bodies such as standards organisations and, although industrial members are involved, they tend to be contributors of expertise and, on occasion, funds. A clear governance structure is often seen as important, as is the openness of the initiatives to a wide range of contribution and transparency on the goals and work of the collaborations.

Tobacco and nicotine industry current collaborations

(a) CORESTA

CORESTA has had a long-standing role as a forum where organisations involved in tobacco and its derived products (manufacturers, suppliers, associations, academia and regulators) can work together on the development and validation of testing methods, tobacco breeding programmes, and Good Agricultural Practices.

Historically, the focus of the testing method development was internal to industry standardisation, such as production quality-related requirements for physical and chemical testing and approaches to evaluating agrochemical residues on tobacco. More recently focus has broadened to include methods for the expanding lists of toxicants that are required to be measured by regulators and machine puffing regimes and liquid/aerosol constituents for electronic cigarettes (24).

CORESTA was founded on resolutions clearly stating "that a close, concrete and permanent cooperation be achieved between the various sectors of study concerning tobacco" and "that the reference systems be unified [and] analytic methods standardised." (25) The association sets out to respond to and resolve the non-competitive issues associated with tobacco and tobacco-derived product manufacture, testing and use.

In 1956 the 24 founding organisations were predominantly European state tobacco monopolies and other government organisations. The current member organisations represent a more diverse range of stakeholders and reflect changes in the industry. Privatisation of state monopolies and subsequent mergers have changed the picture on the cigarette manufacturing side. In addition, tobacco growers, chemical, paper and filter suppliers, and providers of test and production equipment/services have joined CORESTA. Most recently membership growth has been driven by manufacturers and suppliers of e-cigarettes and e-liquids as well as independent testing organisations.

Altogether, the members share knowledge and experience, cooperate in collaborative studies, and supply components, equipment and resources to produce CORESTA Recommended Methods (many of which have become ISO Standards, the process being initiated either by national standards bodies or CORESTA), Guides, Technical Reports, and other material presented at CORESTA meetings and made publicly available on the CORESTA website (www.coresta.org).

CORESTA also has working groups on, for example, the development of standardised methods for *in vitro* toxicological testing of tobacco and nicotine products, and the measurement of biomarkers of exposure to toxicants.

(b) Standards Organisations

ISO Technical Committee TC 126 focuses on tobacco and tobacco products and has 32 participating members, 29 observing members, as well as eight Liaison Organizations, including CORESTA, the European Commission, the Framework Convention Alliance, the World Customs Organization, and the WHO. Over the years, it has published 67 ISO standards, 43 of which are the direct responsibility of TC 126, and is currently working on a further 23. The secretariat is the Deutsches

Institut für Normung (DIN), the German standards organisation, and other members are national standards organisations, including Brazil, India, Russia and China (26).

In 2015 ISO TC 126 established a sub-committee on vapour products in response to a proposal from the Association Française de Normalisation (AFNOR), and is working on issues such as the performance specification for an analytical vaping machine which will underpin analytical methods for the testing of the emissions from e-cigarettes and on an analytical method for determining the major substances in e-liquids. BSI and AFNOR were the first to introduce national standards for vapour products, namely a publicly available specification (27) and three national standards (28) respectively. The work of these national bodies has now progressed into the pan-European standards organisation CEN (European Committee for Standardisation) which has established Technical Committee 437 to develop a range of standards for e-cigarettes. Some aspects of this work may be jointly planned with ISO under the Vienna Agreement (29), thus making the standards international.

Another example, but with quite a different model, was the development of standards for the manufacture and contents of Swedish snus. This began with the Swedish Match company developing and publishing an internal voluntary standard, called Gothiatek® (30), which both covered standards for manufacture and set maximum levels for certain constituents. This standard was later adopted by a broader range of manufacturers of snus under the European Smokeless Tobacco Council (ESTOC) trade association and the approach became known as the ESTOC standard. However, this voluntary quality standard is not adopted or required by regulation.

(c) *In vitro* testing

Various initiatives are seeking to drive the replacement of animal testing with *in vitro* models. One of the leaders in this is the Institute for *In Vitro* Sciences (IIVS) which recently, under an FDA grant for small conferences, hosted a workshop on *in vitro* testing of tobacco and nicotine products with contributions from the tobacco and nicotine industry, *in vitro* model suppliers, academics, and regulators (31).

Similarly, British American Tobacco and Philip Morris International are contributing to the OECD AOP-wiki initiative, working to define two specific adverse outcomes pathways (32).

(d) Consortium on priority ingredients

Responding to Article 6 of the revised European Directive on Tobacco Products (TPD2), an EU expert scientific panel advised that more scientific data was needed on specific additives used in tobacco products, including studies related to toxicity and addictiveness (33). It suggested that tobacco manufacturers could carry out joint studies to collect the required data.

Opportunities for further collaborations

As the breadth of testing required of tobacco and nicotine products increases, there is clearly scope to expand the non-competitive collaborations to cover new fields—whether that be in the development and validation of population models, or the validation of biomarkers of biological effect. Often in these areas, the tobacco and nicotine industry sector will not be the key owners of expertise, so expansion of consortia to include academia and other industrial sectors makes sense if common purpose can be found.

In addition, where new product platforms are emerging, such as in e-cigarettes and tobacco heating products, opportunities exist to develop consensus approaches to testing that can be proposed for formal standardisation through international bodies.

Trust remains critical, so good governance of any initiative with emphasis on openness, transparency and inclusiveness is likely to be important.

Conclusions

In common with many industrial sectors, the tobacco and nicotine industry is actively contributing to the development of standardised test methods and protocols for next-generation products. However, the challenges set by regulators to assess the addictiveness, attractiveness and toxicity of both traditional and new products mean that there is a need for an expansion of collaboration to tackle those needs and emerging sciences. This is likely only to be accepted if there is clear purpose in the initiatives, a common need between industry, regulators and society, and good governance of the effort.

References

1. World Health Organisation, WHO TobLabNet SOP 10 – Standard operating procedure for determination of nicotine and carbon monoxide in mainstream cigarette smoke under intense smoking conditions, Geneva, 2016
2. European Environmental Agency, Poor European test standards understate air pollution from cars, 19th October 2004
3. US FDA, Family Smoking Prevention and Tobacco Control Act, 111 P.L. 31; 123 Stat. 1776; 2009 Enacted H.R. 1256; 111 Enacted H.R. 1256, June 2009
4. European Union, SCENIHR, Attractiveness and Addictiveness of Tobacco Ingredients, 2010

5. CORESTA *in vitro* toxicology testing taskforce, The rationale and strategy for *in vitro* toxicology testing of tobacco smoke, May 2004
6. Fearon *et al*, Nicotine delivery from e-cigarettes, data and learnings from clinical pharmacokinetic studies, Society for Nicotine and Tobacco Research, Annual meeting 2017, available at www.bat-science.com
7. Farsalinos, *et al.* Nicotine absorption from electronic cigarette use: comparison between first and new-generation devices, Sci. Rep., 4, p. 4133, 2014
8. Richard *et al*, ToxCast Chemical Landscape, Pathing the road to 21st Century Toxicology, Chem Res Tox, 29, 1225-1251, 2016
9. International Organisation for Standardisation, www.iso.org/home, accessed 1st August 2017
10. American National Standards Institute, www.ansi.org, accessed 1st August 2017
11. The International Council for Harmonisation of Technical Requirements for Pharmaceuticals for Human Use (ICH), www.ich.org/home, accessed 1st August 2017
12. The International Council for Harmonisation of Technical Requirements for Pharmaceuticals for Human Use (ICH), Integrated addendum to ICH E6 (R1): Guideline for Good Clinical Practice, EC (R2) October 2016
13. Global Food Safety Initiative (GFSI), www.mygfsi.com, accessed 1st August 2017
14. UK British Retail Consortium Food Safety Standard, www.brcglobalstandards.com, accessed 1st August 2017
15. International Food Standard (IFS), www.ifs-certification.com/index.php/en/standards/251-ifs-food-en, accessed 1st August 2017
16. Dementia Consortium, www.dementiaconsortium.org, accessed 1st August 2017
17. World Intellectual Property Organisation, www.wipo.int/research/en/, accessed 1st August 2017
18. American Chemical Society's Green Chemistry Institute Pharmaceutical Roundtable, www.acs.org/content/acs/en/greenchemistry/industry-business/pharmaceutical, accessed 1st August 2017
19. Open Automobile Alliance, www.openautoalliance.net/#about, accessed 1st August 2017
20. Here, HERE gets new investors and plans expansion into China". HERE 360. 27th December, accessed 1st August 2017
21. International Consortium for Innovation and Quality in the Pharmaceutical Sector, iqconsortium.org, accessed 1st August 2017
22. Organisation for Economic Co-operation and Development, Summary report of the 7th meeting of the extended advisory group on molecular screening and toxicogenomics, Paris, ENV/JM/TG/M(2014)4, June 2014

23. Belcastro *et al*, The sbv IMPROVER Systems Toxicology Computational Challenge: Identification of Human and Species-Independent Blood Response Markers as Predictors of Smoking Exposure and Cessation Status, Comp Tox, online 14th July 2017
24. Wright *et al*, Chemical analysis in the regulation of tobacco products, Tobacco Scientists Research Conference, September 2014, available on www.bat-science.com
25. CORESTA, www.coresta.org, accessed 1st August 2017
26. International Organisation for Standardisation, www.iso.org, accessed 1st August 2017
27. British Standards Institute, Vaping products, including electronic cigarettes, e-liquids, e-shisha and directly-related products. Manufacture, importation, testing and labelling, PAS 54115:2015
28. AFNOR, XP D90-300-1, XP D90-300-2 and XP-D90-300-3, 2016
29. ISO, Vienna Agreement, isotc.iso.org/livelink/livelink/fech/2000/2122/3146825/4229629/4230450/4230458/01, accessed 1st August 2017
30. Rutqvist *et al*, Swedish snus and the GothiaTek® standard, Harm Reduct J. May 16;8:11, 2011
31. Institute for *in vitro* sciences, *In Vitro* Exposure Systems and Dosimetry Assessment Tools for Inhaled Tobacco Products, held April 4-6, 2016 in Bethesda, MD
32. Tobacco reporter, BAT and PMI share award, February 13, 2017
33. Hoet *et al*, Recommendations to the European Commission implementing a priority list of additives that should have more stringent reporting requirements: the opinion of the Scientific Committee on Emerging and Newly Identified Health Risks (SCENIHR), Tobacco Control, Special Communication, published online dx.doi.org/10.

FACTORS AFFECTING MEDIA COVERAGE OF TOBACCO RESEARCH NEWS

Cheryl K. Olson
Cheryl K. Olson ScD, LLC
San Carlos, CA USA

Abstract
Scientific publications with perceived relevance to public health or safety are often picked up by the popular press. The accuracy of such coverage by electronic and print media varies substantially. Reviewing examples of coverage of recent scientific publications on health effects of tobacco, nicotine or e-cigarettes reveals patterns in the quality and nature of press coverage, as well as its apparent motivation. This presentation focuses on the process by which journalists decide whether to cover a health-related news story; common points of confusion that affect quality of coverage (*e.g.*, correlation vs. causation, or statistical vs. real-world significance); and ways that researchers might act to improve the likelihood and quality of media coverage of research.

According to the latest U.S. National Science Board survey about public attitudes and understanding about science and technology, medical news is extremely popular; 59% are "very interested" in medical discoveries—more than any other category of news—with just 5% "not at all interested."

One topic of growing interest is research on new electronic nicotine delivery systems. Recent reviews of press coverage of e-cigarettes in the U.S. (2) and U.K. (3) suggest there is "much public confusion surrounding the new technology along with controversy regarding its regulation" (2). While useful, these quantitative studies of reporting provide little insight into why some research attracts more interest than others, how to improve the quality and accuracy of such coverage, and how to credibly address the flaws or limitations of research.

This presentation draws from a review of U.S. mass media coverage produced for Pinney Associates of four recent scientific publications on electronic cigarettes or vaping that received significant coverage in the general press. For space reasons, only two of these articles are addressed here. (Although there may be minor differences, in this document I use the terms "e-cigarettes" and "vaping" devices interchangeably.)

The review of the first article's coverage also looks at the process by which journalists decide whether and how to cover a health-related news story, what makes such coverage more or less effective, and the types of questions that journalists should be asking when assessing such stories.

Story #1: Hidden Formaldehyde

On January 15, 2015, *The New England Journal of Medicine (NEJM)* published a letter titled "Hidden Formaldehyde in E-Cigarette Aerosols"(4) that received global attention. The study measured the output of formaldehyde-containing hemiacetals (a formaldehyde-releasing agent) under laboratory conditions from a commercially available e-liquid when aerosolized by a tank system e-cigarette that used a variable voltage battery.

Formaldehyde is a colorless, flammable, strong-smelling, naturally occurring and common organic compound that can cause irritation of the skin, eyes, nose and throat. In addition to its best-known use as an embalming fluid, formaldehyde is used in many household products including construction materials, glues, paints, fertilizers, pesticides, and permanent press fabrics. It is also a byproduct of combustion, and can be found in cigarette smoke. There is no doubt about its cancer-causing properties given sufficient exposure. In 2014, the National Toxicology Program listed formaldehyde as a known human carcinogen (5).

The "hook" for the news media coverage of the letter was that it appeared to show new links between e-cigarettes and cancer that drew parallels with the well-established link between tobacco cigarettes and cancer. But did the research described in the letter really do that? And did the press report on that research accurately?

Concerns about the *NEJM* study are manifold and include such fundamental issues as validity (Am I measuring what I think I'm measuring?), reliability (Am I doing a good job of measuring the key variables?), and applicability (Is my model an appropriate representation of real-world situations?). For example:

- Are formaldehyde and "formaldehyde-releasing agents" equivalents when it comes to cancer risk?
- Was the voltage the researchers used in their experimental apparatus (5.0V) a realistic stand-in for real-world vaping, especially since formaldehyde-releasing agents were not detected at the more-typical 3.3V level?
- Are the reference levels for tobacco-generated formaldehyde valid and reliable, especially since they do not overlap at one standard deviation around their respective center points?
- Is the change in relative risk (if it exists) misleading because the absolute risk is so low?
- Is voltage of the delivery device the right independent variable? Power may be more useful since it's more likely to be associated with aerosolized formaldehyde.

Media Headlines for *NEJM* Story

At first glance—and especially to a non-scientist—the research seemed to show that a person using an e-cigarette would be exposed to approximately five times the amount of formaldehyde (14.4 mg/day vs. 3 mg/day) as a one-pack-per-day tobacco cigarette smoker, thereby challenging the claims that e-cigarettes are "safer" than tobacco cigarettes. In many cases that was, indeed, the headline.

- "E-cigarettes Can Produce More Formaldehyde Than Regular Cigarettes, Study Says" - *Los Angeles Times*, 1/21/15 (6).
- "Study Links E-Cigarettes to Formaldehyde, Cancer Risk; Research Found E- Cigarettes Produce New Type of Formaldehyde When Heating Nicotine-Laced Liquid" - *Wall Street Journal*, 1/21/15 (7).
- "E-Cigarettes Can Churn Out High Levels Of Formaldehyde"—National Public Radio, 1/21/15 (8).
- "E-cigarette Vapor Filled With Cancer-Causing Chemicals, Researchers Say" - CBS News, 1/22/15 (9).
- "Before You Vape: High levels of Formaldehyde Hidden in E-Cigs"—NBC News, 1/21/15 (10).

While those headlines were dramatic, some of the coverage was more nuanced. Many of the pieces featured a statement by Greg Conley, president of the American Vaping Association, calling into question the study's use of high voltage settings, *e.g.* "When the vapor device was used at the realistic setting of 3.7 volts, levels of formaldehyde were similar to the trace levels that are released from an FDA-approved [smoking-cessation] inhaler. However, when the researchers increased the voltage to 5 volts and continued to have their machine take three- to four-second puffs, this caused extreme overheating and the production of formaldehyde... These are not settings that real-life vapers actually use, as dry puffs are harsh and unpleasant." (9)

Media Coverage—CBS This Morning

This combination of an alarmist headline and/or "lede" (first paragraph of a news story) with a scientific perspective on the numbers and risks can be clearly seen in a roundtable set piece broadcast January 21, 2016 on *CBS This Morning* (11).

This story is a useful exemplar because it combines some of the things that comprise excellent scientific reporting and others that demonstrate inaccurate and even dangerous reporting.

News anchor Norah O'Donnell begins:

> *There's a new warning this morning that e-cigarettes may not be as safe as they seem. The New England Journal of Medicine says that e-cigarette users are 5 to 15 times more likely to get formaldehyde-related cancers than long-term smokers. Formaldehyde is the toxic chemical found in the devices.*

Dr. Holly Phillips is with us. Dr. Holly, good morning.

Formaldehyde – we think about lab in high school when we were dissecting frogs. That's in e-cigarettes? How dangerous is it?

At this point, Dr. Phillips tries to put the findings and the fears into perspective.

Formaldehyde is a really common chemical. It's not only in frogs in biology labs. It's in permanent press fabrics. It's in glues. And most importantly, it's in regular cigarettes. When you smoke a regular cigarette, you inhale the formaldehyde, which we know can cause cancer.

In the study, researchers took a high-power form of e-cigarette—it's called a tank system—and they basically created a vapor in a lab and studied it. They found that when you heated the vapor up at a low voltage, 3.3 volts, there was very little formaldehyde in it. In fact, none. But when you heated the vapor up at 5 volts, there was a lot of formaldehyde. In fact, two-and-a-half times the amount of formaldehyde that you would get if you smoked a regular cigarette.

News anchor Gayle King then introduces the industry's perspective:

Well, the American Vaping Association—no surprise—is not happy. They say that the study is flawed. How come?

Dr. Holly Phillips:

Not only the Vaping Association, but a number of critics have come out against this paper that was published in The New England Journal. Basically, they're saying what happens in the lab doesn't necessarily happen in real life. And even though they heated up the vapor to a very high voltage (sic) in the lab, real-life e-cigarette users wouldn't do that.

The vapor would taste terrible. It would basically overheat their device. And they would not enjoy the smoking. They would usually smoke at a lower voltage, which has less formaldehyde.

News anchor Jeff Glor responds:

So they're saying you'd have to crank up the e-cigarette in some way. That most e-cigarettes, you can't even do it with to achieve these levels.

Dr. Holly Phillips:

Exactly. They used a specific high-power e-cigarette called a tank system, and so most people wouldn't be exposed in that way.

Norah O'Donnell:

When are we really going to know if they're safe right now?

Dr. Holly Phillips:

> *Theoretically, they should be safer. We know that the main harm from cigarettes isn't the nicotine, even though that's an addictive substance. It's the 4,000 chemicals, 60 of which we know are carcinogens, that come from burning tobacco.*
>
> *E-cigarettes don't burn tobacco. So in theory they should be safer. But as we see today, the jury is still very much out on that.*

Jeff Glor:

> *Holly Phillips. Thanks very much.*

Deciding to Cover the Story: Points of Contact

Before dissecting this story, let's take a look at how it likely got on the air at all. Without that process, there would be nothing to examine. Stories like this are usually triggered by a news release—in this case, almost certainly from Portland State University (12) where the research took place. (Since the publication was a letter, not a refereed research article, it's doubtful that *NEJM* put it in its weekly outreach to journalists.) Because it was information published in a scientific journal, it would have been embargoed until the evening before the publication date of the print edition of that journal. Although they could not report on it until the embargo date, journalists would have had access to pre-publication copies of the letter, including contact information for the researchers.

Approximately one week before broadcast, that press release would have been received by or made its way to a producer at *CBS This Morning*, who would have pitched it for the "budget" (rundown of stories) of the first show to air after the embargo. The decision made by the producers during that budget meeting would be (1) whether to cover the story at all, (2) how to cover the story, *e.g.*, a "reader," field package, live satellite interview with a researcher, roundtable discussion, etc., and (3) how much time to allocate to it. The producers decided to go with a mid-level approach: a roundtable discussion with one of their in-house medical experts. They allocated two minutes and 30 seconds for the story, which is consistent with that mid-level ranking for a morning news program.

It's important to note that the three news anchors discussing this story have no training or expertise in science or medicine. (There are times when, from the journalists' perspective, this is not a problem and may even give them an advantage since it allows them to view news stories from their audience's perspective, acting as their surrogate.) Nor is it likely that any of the show or segment producers have any significant education in medicine or science, or any experience as a beat reporter/producer covering health issues or science; almost all producers have a "general assignment" background. They decided to bring in Holly Phillips, M.D., a board-certified internist and part-time journalist at CBS specializing in health and medicine.

In most cases, the graphics are ordered and assembled by the segment producer, who also writes the lede (read by Norah O'Donnell) and obtains background information and quotes such as the one from the American Vaping Association. The segment producer then briefs the journalist, who may do additional background research, and the anchors, including sharing descriptions of the proposed graphics so that they can allude to the information on them in a timely manner. In a roundtable such as this, everything after the lede is ad libbed. All of the participants, especially Dr. Phillips, will have notes on the key points they'd like to make during the ad lib discussion so that they can make use of the graphics.

It's critical to understand this process and its correlates among print, online and radio outlets to appreciate that there are multiple points of potential contact that can be used to make the coverage of science-related news more accurate. Also, many of the people involved in screening and structuring health and science stories in all media have no significant training in those fields, and thereby have few ways to analyze the validity, reliability or importance of the data presented to them.

Analyzing the CBS News Story

The CBS story begins with an implicit link to the long-term misleading of the public and the press by the tobacco industry regarding cancer risks, following which that industry essentially lost all of its credibility on health-related matters. The thought that another industry, whether it's automobile airbags, pharmaceuticals, or laminated flooring, might be misleading the public on the health risks of its products is red meat to most journalists. Such temptation, combined with a lack of sophistication in science and statistics, can easily lead those journalists—even the ones with the best of intentions—and the public astray.

That implicit link helps explain the overall tone of the piece, including the constant bottom-third graphic that reads "Vapor Danger. Study: Hidden Carcinogen Found in E- Cigarettes." That tone sets the stage for the ominous lede:

> The New England Journal of Medicine *says that e-cigarette users are 5 to 15 times more likely to get formaldehyde-related cancers than long-term smokers.*

Such phrasing implies (inaccurately) to viewers that an august body of physicians (*NEJM*) has determined that e-cigarettes are inherently life-threatening. In reality, the journal simply published a peer-reviewed letter, not a peer-reviewed article, in which the authors describe an experiment they did and present what they see as its implications.

While it's useful and appropriate to consider the source of a story (*e.g.*, *NEJM* vs. a corporate or university public relations office) in an initial screening, most journalists do not appreciate or pay attention to the subtle differences between types of professional journals and types of pieces within those journals. Nor do they appreciate the need for independent replication of findings before they should be generally accepted.

Key Questions Journalists Should Ask

The producers and anchors should not be expected, by themselves, to know which questions to ask when assessing these types of data. But they should be open to and able to ask those questions when provided guidance by a credible source either inside or outside their newsroom. Those key questions include:

- **Validity:** Are the researchers measuring what they think they're measuring? (In this case, is the concentration of formaldehyde the same as concentration of formaldehyde-releasing agents?)
- **Reliability:** How good of a job are they doing at making those critical measurements? Is the experiment structured in such a way that others would get the same results using the same protocol? Would the results stay the same if the same researchers repeated the study a month from now?
- **Real-world modeling:** Is the research design a realistic way of simulating what real people who vape do? Is 3ml/day a reasonable estimate of consumption among vapers? Are the authors leaping to unfounded conclusions?
- **Relative risks vs. absolute risks:** Which are we describing? What sounds like a dramatic change in relative risk can be a trivial change in absolute risk. In this study, while the relative risk goes up by a factor of 5 (or even 15 in some models), the absolute risk is still extremely low. Indeed, in its response to the study, the American Council on Science and Health (13) states, "Biologically, as scary as it sounds (embalming fluid) formaldehyde is actually a very weak carcinogen, with only a slightly increased chance of cancer among even highly-exposed workers over an entire lifetime." In addition, the majority of tobacco cigarette-related cancers are not caused by formaldehyde, but by other carcinogenic compounds that are not present in e-cigarettes. So while the risk of that small subset of cancers may go up, the overall theoretical risk of smoking-related cancers—what the journalists' audience should be concerned about—would go down dramatically! This also has implications for determining statistical vs. clinical significance.

- **Clinical vs. statistical significance:** Statistical significance, which is often a major component in whether an academic research paper is published, is a matter of calculating the odds that the difference on some variable between two or more groups is due to something other than chance or random variation. Clinical significance in a health-related study is a matter of whether the difference between those variables has an important effect on people's lives. The two are not always the same.
- **Frame of reference**: Are the risks and benefits couched in appropriate terms with respect to the reference groups? For example, are we comparing cancer rates among vapers to those among tobacco cigarette smokers or to cancer rates among lifetime non-smokers? Are the exposure statistics for tobacco cigarettes and e-cigarettes comparable? In this study, they're comparing e-cigarette users to tobacco cigarette smokers, which is a valid reference group since e-cigarettes are promoted as a way to quit smoking. But, as noted above, the conclusion about theoretical health risks with respect to that reference group did not cite the overall decline in all smoking-related cancers vs. the increase in formaldehyde-related cancers.
- **Clear definitions:** Are both the technical and general terms used by the researchers clearly and consistently defined? For example, what constitutes a "daily user" of an e-cigarette? What about a "regular user" or an "experimenter?" Do other researchers use those terms in the same way?
- **Appropriate statistics**: Are the researchers using appropriate mathematical techniques to analyze their data? Have others used those same techniques in similar studies?
- **Vested interests:** Do the authors have any vested interests or conflicts of interest that might influence how they gather, interpret or present their results?

Health and science beat reporters—especially those who have professional training in the sciences—routinely ask those and other questions. Unfortunately, most general assignment reporters and producers do not analyze stories and story pitches this way. The *CBS This Morning* story is a case in point. It begins with an ominous statement that sets the viewers' expectations by using the words "warning," "cancer" and "toxic":

> *There's a new warning this morning that e-cigarettes may not be as safe as they seem. The* New England Journal of Medicine *says that e-cigarette users are 5 to 15 times more likely to get formaldehyde-related cancers than long-term smokers. Formaldehyde is the toxic chemical found in the devices.*

The tone of the piece is reinforced by a full-screen graphic.

Dr. Phillips's initial statement on-air takes a different and more-scientific approach in an attempt to put the audience's exposure to formaldehyde into perspective and to rein in the credulousness of the anchors:

> *Formaldehyde is a really common chemical. It's not only in frogs in biology labs. It's in permanent press fabrics. It's in glues. And most importantly, it's in regular cigarettes.*

This temporarily shifts the tone of the piece away from melodrama and begins to explore some of the key questions mentioned above.

Gayle King pivots to the industry comment while expressing her skepticism:

> *Well, the American Vaping Association—no surprise—is not happy. They say that the study is flawed.*

This cynicism about industry sources has profound implications for determining who has credibility on health and science matters with journalists.

Dr. Phillips attempts to buttress the concerns about the researchers' apparent design flaws:

> *Not only the Vaping Association, but a number of critics have come out against this paper that was published in The New England Journal. Basically, they're saying what happens in the lab doesn't necessarily happen in real life. And even though they heated up the vapor to a very high voltage (sic) in the lab, real-life e-cigarette users wouldn't do that.*
>
> *The vapor would taste terrible. It would basically overheat their device. And they would not enjoy the smoking. They would usually smoke at a lower voltage, which has less formaldehyde.*

The unnamed critics that she cites are amorphous and therefore lack credibility. The audience does not know if these are reputable scientists, defensive vapers, or public relations flacks.

Jeff Glor responds by recapitulating what Dr. Phillips is saying about the study's design flaws:

> *So they're saying you'd have to crank up the e-cigarette in some way. That most e-cigarettes, you can't even do it with to achieve these levels.*

This is a key point in covering this story: the real-world situation being modeled is unrealistic. Therefore, the conclusions about cancer risk do not appear to be valid. Since we don't have measurements at voltages other than 3.3V and 5.0V, and we don't know from this study what voltage a typical vaper uses, we simply don't know what the risk from vaping for formaldehyde-related cancers is. Unfortunately, this key point is missed in the exchange at the end of the story:

Norah O'Donnell:

> *When are we really going to know if they're safe right now?*

Although it's awkwardly phrased, this is the most important question to the viewers: What light does this study shed on the absolute dangers of using e-cigarettes and their relative dangers when compared to tobacco cigarettes? Dr. Phillips responds:

> *Theoretically, they should be safer. We know that the main harm from cigarettes isn't the nicotine, even though that's an addictive substance. It's the 4,000 chemicals, 60 of which we know are carcinogens, that come from burning tobacco.*
>
> *E-cigarettes don't burn tobacco. So in theory they should be safer. But as we see today, the jury is still very much out on that.*

Everything in her comments is accurate until that last line. The research they reported on did not confirm that "the jury is still very much out" on e-cigarette safety. It was a flawed study that was treated by the anchors and the graphics as if it signaled a dramatic increase in cancer risk when it did nothing of the sort.

Media Coverage – National Public Radio

National Public Radio's *Morning Edition* ran a piece by Rob Stein on the *NEJM* formaldehyde story. (14) Stein is the senior editor on that network's science desk who's spent most of his career covering medicine and science for newspapers and broadcast media. While he was more circumspect, he also indulged in a bit of melodrama and missed some key issues.

Renee Montagne, host:

> *Here's something to think about when taking a puff of an e-cigarette—formaldehyde. New research is raising more concern about the safety of electronic cigarettes, finding the vapor they produce contains more formaldehyde than previously reported. NPR's Rob Stein has this story.*

Rob Stein:

> *E-cigarettes work by heating up a liquid that contains nicotine. That makes a vapor that users inhale. It's called vaping. E-cigarettes are generally considered safer than tobacco cigarettes, but David Peyton of Portland State University and his colleagues decided to take a closer look at what's in that vapor.*

Dr. David Peyton:

We simulated vaping by drawing the vapor—the aerosol—into a syringe, sort of simulating the lungs... (e)

To our surprise, we found a form of formaldehyde in e-cigarette vapor.

Rob Stein:

A form that might make it easier for formaldehyde to slip into someone's lungs. And they didn't just find a little formaldehyde; they found a lot.

Dr. David Peyton:

We found this form of formaldehyde at significantly higher concentrations than even regular cigarettes—between five and 15 fold higher concentration of formaldehyde than in cigarettes.

Rob Stein:

And formaldehyde can be nasty.

Dr. David Peyton:

Long-term exposure is recognized as contributing to lung cancer and so we would like to minimize, to the extent one can, contact with formaldehyde, especially delicate tissues like lungs.

Stein does a good job of quickly explaining how e-cigarettes work and that they're generally considered to be safer than tobacco cigarettes. He reached out to one of the study's authors to get greater insight. However, that author's statement, "To our surprise, we found a form of formaldehyde in e-cigarette vapor" strains credibility given the researchers' backgrounds in chemistry. While the quantity of formaldehyde released at different voltages may have been surprising, its presence should not have been given the composition of the e-liquid they were heating. It makes it sound all the more frightening, as if formaldehyde was a hidden and unexpected threat.

Rob Stein:

For their part, companies that make e-cigarettes are dismissing the study. Gregory Conley of the American Vaping Association says the researchers only found formaldehyde when the e-cigarettes were cranked up really high.

Gregory Conley:

> *No real-life human is ever going to vape at that setting throughout the day because after a couple puffs, they'd be unable to puff anymore. They would take the vapor product and take a puff for one second and it would burn. Not burn like a third-degree burn, but it would feel extremely unpleasant in your lungs.*

By including only an industry representative, Stein is setting up the story as "science vs. industry," ignoring some of the serious questions other scientific researchers have about the experiment and the interpretation of its results. This skews the information in favor of the *NEJM* letter's authors, even if that's not warranted.

Conley then questions the applicability of the experimental design to real-world vaping and follows that soundbite with his charred steak analogy:

> *"I can take a steak and I can cook it on the grill for the next 18 hours, and that steak will be absolutely chock-full of carcinogens. But the steak will also be charcoal, so no one will eat it."*

Rob Stein:

> *Peyton, the researcher, acknowledges that he found no formaldehyde when the e-cigarettes were set low. But he says he thinks plenty of people are using the high settings.*

Dr. David Peyton:

> *As I walk around town and look at people using these electronic cigarette devices, it's not difficult to tell what sort of setting they're using. You can see how much of the aerosol they're blowing out. It's not small amounts. It's pretty clear to me that at least some of the users are using the high levels.*

Rob Stein:

> *Peyton hopes the government will limit the sale of these devices, especially to kids. The Food and Drug Administration is in the final stages of trying to decide how strictly the agency will regulate electronic cigarettes.*
>
> *Rob Stein, NPR News.*

Peyton's claim that he can "tell what sort of setting [vapers are] using" is unsubstantiated and questionable. It should be challenged.

Media Coverage – Reuters

Toni Clarke, a biotechnology reporter for Reuters, did an excellent job of covering the story on January 21, 2015 (15). She begins:

> **Ramping up e-cigarette voltage produces more formaldehyde—study**
>
> *People who smoke high-voltage e-cigarettes have greater exposure to formaldehyde, a suspected carcinogen, than those who keep the voltage low, according to a study published in the New England Journal of Medicine on Wednesday.*
>
> *The study, which critics say is misleading and lacks context, is the latest contribution to a debate on the safety of e-cigarettes that so far has yielded little long-term data, though most experts believe they are less toxic than combustible cigarettes...*

After explaining the underlying science and the research design, Clarke then does something basic that the other reporters cited above failed to do: she speaks with both a researcher who wrote the *NEJM* letter and other researchers in the field to clarify some issues. Her interview with coauthor Dr. James Pankow brings to light some very important limits to interpreting the findings that were missed by other press reports.

> *It is not known exactly where formaldehyde contained in hemiacetals gets deposited in the body or whether it is similarly toxic, said James Pankow, one of the study's authors.*
>
> *"There has never been a cancer study with hemiacetals," Pankow said in an interview.*
>
> *Absent such a study, the authors estimated the formaldehyde-related cancer risk associated with e-cigarettes by extrapolating from data on formaldehyde in cigarettes.*

This is the essence of strong science reporting: clarity not only about the research but also about its limits and implications. Hemiacetals may not be a valid surrogate for aerosolized formaldehyde. Extrapolated risk may not translate from one chemical to the other. Relative overall cancer or health risk may be more important than absolute formaldehyde-related cancer risk.

What's particularly powerful is that these challenges to the interpretation of the results of the study come not from industry representatives, who would be perceived to have a financial interest in a particular outcome, but from other researchers who are professional peers of the authors: Dr. Jed Rose, director of the Center for Smoking Cessation at Duke University Medical Center, Dr. Neal Benowitz, a nicotine expert at the University of California, San Francisco, and Dr. David Abrams, executive director of the Schroeder Institute for Tobacco Research and Policy Studies at the anti-tobacco group Legacy. Each of these commentators brings with him the imprimatur of his institution, giving greater weight to their statements and therefore greater credibility and usefulness to the reporter's piece.

Story #2: Flavored E-Cigarettes and Youth

On November 7, 2016, the journal *Pediatrics* published a study by Hongying Dai and Jianqiang Hao (16) on "Flavored Electronic Cigarette Use and Smoking Among Youth." Their paper used data from the 2014 National Youth Tobacco Survey (NYTS) (17) an annual questionnaire given to students in grades 6-12 who comprise a representative sample of students that age throughout the country. (Note that children who drop out of school in those grades are significantly more likely to use tobacco products than children who are still in school, so the sample may not be representative of the overall population when it comes to smoking and vaping.)

> *A logistic regression model was used to assess whether flavored e-cigarette use was associated with (1) intention to initiate cigarette use among never-smoking youth (n=16,471), (2) intention to quit tobacco use among current-smoking youth (n=1,338), and (3) perception of tobacco's danger among all respondents (n=21,491).* (16)

Logistic regression is a way of combining several independent variables in an attempt to predict a dependent variable that's categorical, *e.g.*, alive/dead, smokes/ doesn't smoke, diagnosis of Disease A/Disease B/Disease C.

> *Concern is growing that widespread availability of flavored e-cigarettes will increase the use of e-cigarette products by youth and will thus reinforce the acceptability of vaping behavior. The normalization of e-cigarette use among youth could also lead to e-cigarettes becoming a gateway for future smoking, marking a setback in the decades-long antismoking battle.* (16)

This is a "hot button" issue for journalists as well as public health advocates. If we embrace harm reduction by promoting e-cigarettes as a way of quitting tobacco cigarettes, are we unintentionally luring children who otherwise wouldn't smoke tobacco to start vaping, thereby increasing their health risks? Or are children (or most of those children) who vape the ones who otherwise would have taken up smoking tobacco cigarettes, and are thereby reducing their health risks—although not as much as they would have by abstaining completely?

Media Coverage – Reuters

Kathryn Doyle, a health reporter for Reuters, picked up on this research in an article published on November 7, 2016 titled "Kids Who Use Flavored E-cigs More Likely to Want to Try Cigarettes." (18) It begins:

> *U. S. kids who use flavored e-cigarettes more often intend to start smoking traditional cigarettes than kids who did not use flavored vapes, according to a 2014 national survey.*

A few paragraphs down, however, she interviews another researcher who questions the conclusions implied by the headline.

> *"This study does not show e-cigarettes are definitely a gateway to smoking cigarettes," said Shu-Hong Zhu, principal investigator of the California Smoker's Helpline, who was not part of the new study. "It shows that students who tried e-cig are more interested in cigarettes than students who have not tried e-cig," but is not sure evidence of a gateway from one to the other.*

In other words, the Dai and Hao study confuses correlation with causation. It may be, for example, that students who are interested in smoking tobacco cigarettes are more likely than non-smokers to want to try e-cigarettes for reasons that influence both decisions.

Media Coverage – HealthDay News

On November 7, 2016, HealthDay News, a syndication service specializing in medical stories, published a piece by Steven Reinberg titled "Flavored E-Cigarettes May Entice Teens to Smoke." (19) It leads with:

> *Fruit- or candy-flavored electronic cigarettes may entice American teens to start smoking tobacco, a new study suggests.*
>
> *... "Due to a proliferation of e-cigarette flavors on the market, flavored e-cigarette use among youth in the U.S. has increased significantly," study author Hongying Dai said. She's an associate professor of health services and outcomes at Children's Mercy Hospital in Kansas City.*

That first statement is a confusion of correlation with causation. So is the second. (Do we know that the number of youth trying e-cigarettes is "due to a proliferation of e-cigarette flavors?" Or were the range of flavors developed to capture segments of the e-cigarette using population?)

The second statement is also a distortion due to confusing absolute numbers with relative numbers or ratios. The modern e-cigarette was invented in 2004 (20). It has only been marketed nationally for a few years. As with any successful new product, we would expect its use to increase significantly because the denominator (the absolute number of youth vaping) in that ratio starts at a low number. Since its base rate (reference point) is so small, any growth is exaggerated when expressed in those terms.

For example, the US Centers for Disease Control and Prevention issued a press release (21) in April 2015 with the headline "E-cigarette Use Triples Among Middle and High School Students in Just One Year."

On its own, that sounds terrifying. But a few sentences into the release we learn that the dramatic percentage jump is due to two things: a broad definition of e-cigarette use and a low reference point. Current use was defined as at least one puff in the previous 30 days, thereby including those students who experimented with and rejected e-cigarettes. Among middle school students, the "current use" rate went from 1.1 percent in 2013 to 3.9 percent in 2014.

There's another likely reason for the dramatic jump: the survey questions changed. For example, in 2013 the NYTS question about use of "other forms of tobacco" during the previous 30 days was:

> 37. In the **past 30 days**, which of the following products have you used on **at least one day? (You can CHOOSE ONE ANSWER or MORE THAN ONE ANSWER)**
> A) Roll-your-own cigarettes
> B) Flavored cigarettes, such as Camel Crush
> C) Bidis (small brown cigarettes wrapped in a leaf)
> D) Clove cigars (kreteks)
> E) Flavored little cigars (such as mint, clove, spice, alcohol (wine, cognac), candy, fruit, chocolate, or other sweets)
> F) Smoking tobacco from a hookah or a waterpipe
> G) Snus, such as Camel or Marlboro Snus
> H) Dissolvable tobacco products, such as Ariva, Stonewall, Camel orbs, Camel sticks, or Camel strips
> I) Electronic Cigarettes or E-cigarettes, such as Ruyan or NJOY
> J) Some other new tobacco products not listed here
> K) I have not used any of the products listed above or any new tobacco product in the last 30 days

In 2014, that question was changed to:

> 35. In the **past 30 days**, on how many days did you use electronic cigarettes or e-cigarettes such as Blu, 21st Century Smoke, or NJOY?
> A) 0 days
> B) 1 or 2 days
> C) 3 to 5 days
> D) 6 to 9 days
> E) 10 to 19 days
> F) 20 to 29 days
> G) All 30 days

Indeed, the CDC admits that this might be the case in an article in *Morbidity and Mortality Weekly Report* (22) about the survey's methodology:

> *Changes between 2013 and 2014 in the wording and placement of questions about the use of e-cigarettes, hookahs, and tobacco pipes might have had an impact on reported use of these products.*

A quote from Greg Conley in the HealthDay News piece called Dai and Hao's research design into question for yet another reason:

> *"In this study, any teen who answered 'probably not' when asked if he or she thought they would smoke a cigarette in the next year was marked as 'intending to smoke,'" said Gregory Conley, a spokesman for the American Vaping Association.*
>
> *Only a small number of participants answered "definitely" to the cigarette smoking question, Conley said. "So in order to give themselves statistical power, the researchers enlarge the category to include 'probably not,'" he said.* (19)

This is a recurring issue in scientific research. Even though their value to science may be the same, journal editors are much more likely to publish articles in which the findings show a statistically significant difference of some sort. That puts pressure on researchers to design their studies to show some type of statistical significance, even if it's an inappropriate or suboptimal design.

The piece ends with a quote from Dr. Stanton Glanz of the University of California at San Francisco, who is a strong advocate of the abstinence approach.

> *"This reality makes the fact that the Obama White House dropped regulation of flavors, which especially appeal to kids, from the FDA's recent rule taking jurisdiction over e-cigarettes worrisome," Glantz said. "The effect of this deletion will delay regulation of flavors in e-cigarettes by years, leading more kids to get addicted to nicotine," he said.* (19)

This once again conflates the supposed dangers of nicotine with the well-established and far-worse dangers of smoking tobacco.

Coverage of this *Pediatrics* article was an interesting case study in the translation of an academic paper into journalism for the lay public. The intersection of health risks and teenagers is always a provocative topic. Fruity flavors, perhaps because of their association with even younger children, would normally make this subject matter catnip to reporters and editors.

Given the press response to earlier studies of e-cigarettes as "gateway drugs" to tobacco cigarettes, this research received comparatively little coverage—probably due to its publication online on November 7, 2016, one day before the U.S. presidential election. The study was simply overshadowed by more pressing issues.

Common Errors In News Coverage

In sum, while some of the reporting on tobacco research was clear, practical and evocative, other media interpretations had repeated and predictable patterns of distortions, poor interpretation and missed opportunities. These problems included:

- **Alarmism.** With the *NEJM* study coverage, many reporters latched onto the implicit and explicit dangers of aerosolized formaldehyde without putting those dangers into a useful (and calming) context. Such key words as "cancer" and "cigarettes" appear to have contributed to this, as did the association of formaldehyde with embalming fluid, and therefore death. Very few reporters paid close attention to the potentially positive finding of no detectable level of formaldehyde in the low-voltage condition, which is the setting used by many vapers, especially those who use single-voltage systems.
- **Credulity.** Most reporters either didn't have the knowledge or felt it inappropriate to question the findings of the research. Non-specialists also showed little understanding of the nature of scientific research, viewing the results of a single experiment or survey as a kind of immutable truth rather than a step toward greater understanding that should be challenged to determine its robustness. Such constant challenging is at the heart of science and drives progress.
- **Poor choice of interview subjects.** Many journalists simply reached out to a small number of industry spokespeople or public health advocates. The best reporters not only reached out to the original researchers for clarification of details and interpretation, they also contacted other researchers who specialized in this field but had no financial ties to industry. Such scientists are seen as experts and credible peers of the original researchers, and their concerns or support are given greater weight by readers/viewers/listeners.
- **Failure to acknowledge constraints of the study.** Most studies are narrowly focused, but the reporting on them often presented them as broad or generalizable. The journalists' interpretation of the study's results and implications often went well beyond the actual data.
- **Not specifying reference groups.** In multiple reports, words like "healthier" and "riskier" were used without specifying healthier than whom or riskier than what. There was also a confusion of absolute risk with relative risk.
- **Hidden motives.** This is often thought of in terms of industry spokespeople wanting to promote their products and downplay any risks, or researchers wanting to burnish their reputations or get tenure at their university. But there are others, including conflicts of interests by journalists or news outlets. For example, the local news website CapeCod.com ran a feature

story on November 23, 2015 titled "Do E-cigarettes Help You Quit Smoking?" (23) that linked to the *NEJM* article.

Although it's not obvious at first glance, the piece is simply a reprint of a press release issued by Cape Cod Health News, which is owned by Cape Cod Healthcare (CCHC). CCHC owns two hospitals and manages the practices of 450 physicians in Cape Cod, Massachusetts. Thus, the main purpose of Cape Cod Health News is to provide a venue for promoting its physicians and services, not to provide the most accurate health information. This is not made clear to the reader. But it does explain why the two physicians (a cardiologist and an emergency physician) who commented on the study were not appropriately qualified experts in the fields in which they expressed their opinions. They were selected because of the CCHC-employed writer's mandate to burnish the image of the CCHC hospitals.

- **Beat and general assignment reporters tend to cover these stories differently.** Those journalists who have expertise in science or medicine tend to view research results more skeptically and to ask questions that challenge that research. General assignment reporters tend not to challenge researchers' methods or conclusions.

The news media want to get it right. There's no long-term incentive for journalists to distort their coverage of science stories. But as demonstrated in the examples above, news media of all types have difficulty covering complex health and science-related topics accurately and insightfully for a variety of reasons. Here are some key points to keep in mind:

- **Many of the reporters/producers/editors charged with interpreting these stories have little or no formal training or experience in science and mathematics,** which may result in their being inappropriately credulous or alarmist, and which makes it more difficult for them to put research findings into an appropriate perspective. Those who have significant professional experience and/or academic training in health or science tend to approach covering scientific research differently than those who are general assignment reporters and cover these fields infrequently.
- **Stories tied to smoking—and especially tied to smoking and cancer—are tainted by the history of misrepresentation of scientific data on risk to the media and to the public by the tobacco industry.** This predisposes journalists and the public to assume that all smoking- and vaping-related products have similar health risks, when that's not true. People tend to act based on their emotions, not on their logical thoughts. This is a constant challenge in the coverage of health-related topics.
- **Industry representatives and spokespeople have little credibility when it comes to discussing the health risks of their products,** even when they have independently gathered data to back up their claims. They are called upon to comment because it's easy to reach them and it provides support

for a claim of balanced coverage. There's no resource not controlled by the industry that helps reporters make sense of this research.

- **Protecting children from new technologies, whether it's video games or e-cigarettes, because of their unsubstantiated but purported dangers has shown itself to be a powerful political bandwagon.** No one wants to be accused of being against children's safety. This makes a practical *harm reduction* approach (*e.g.*, getting teenagers who smoke tobacco to switch to vaping) less emotionally attractive than an impractical *harm elimination* approach (*e.g.*, get teenagers who smoke tobacco to quit completely). Many of those who promote harm elimination position that approach as if it's an easily achievable state, which it is not. In fact, it's likely impossible. Adolescents are biologically driven to take risks as a way of testing adulthood and establishing their independence (24).
- **There are times when media outlets confuse accuracy with the appearance of balance.** In the *CBS Morning News* story, for example, we have no doubt that the journalists strove for accuracy. The problem was that most of them didn't have the expertise and perspective to know what questions to ask that would help explicate the study. In their efforts to achieve accuracy, they looked for balance. In this case, it meant getting a comment from a group that was bound to disagree with the researchers' conclusions: a representative of the vaping industry. The grudging quality of this attempt at balance is shown by Gayle King's statement, "Well, the American Vaping Association—no surprise—is not happy."

  Obtaining this type of industry comment does not improve the accuracy of the science news story in part because *the source of that information is inherently perceived as tainted, even when the information is independently gathered and accurate.* Although the formaldehyde study was published as a letter to the editor and not as a full peer-reviewed article, it carried with it the imprimatur and gravitas of *The New England Journal of Medicine.* A spokesperson for an industry group or corporation has no such weight behind his or her words.
- **Even when journalists try to achieve balance, they sometimes do so by creating false equivalencies.** This is most obvious when journalists covering emotionally charged issues such as evolution, which has been agreed upon by the vast, vast majority of scientists for decades, feel obligated to obtain a countering quotation from a non-scientist who claims that the Earth is only 6,000 years old and was created in seven days. Such false equivalencies run counter to the scientific method.

  We should be concerned about false equivalencies involving the health risks of vaping. We should especially be sensitive to the differences between absolute risk and relative risk. No one can legitimately claim that vaping is risk-free. There is significant evidence, however, that vaping carries fewer health risks and different health risks than smoking tobacco. It's unrealistic—a false equivalency—to assume that no one will take any type of drug that alters their state of consciousness. Yet that position of total abstinence is often used unquestioningly as the reference point for assessing potential risk of a substance or product.

- **There are multiple points of contact within news media that can be used to improve the accuracy of news coverage of scientific research.** Using the example of the formaldehyde related cancer story reported by *CBS This Morning*, the program producer, the segment producer and the beat reporter each had considerable influence in how that information was presented. In other media, the contact person might be an assignment editor or a health editor.

Conclusion

There are several things that those involved in noncombustible nicotine delivery products can do to improve the quality of science news reporting in this area:

- **Focus on improving accuracy of reporting rather than just burnishing image or boosting sales.** Remember the lessons learned by the tobacco industry when it undermined its own credibility by hiding and distorting data. If you're perceived as simply pushing a self-dealing agenda, you will not be listened to.

  There are times when research studies will point to increased risks, problems and other negative effects. These are to be expected; no product or procedure is perfectly safe and effective. But industry spokespeople and researchers must maintain their credibility with journalists if they are to help put these stories into perspective. Let's say, for example, that a new and well-designed study finds that the theoretical risk of formaldehyde-related cancers among long-term e-cigarette users is three times the theoretical risk of formaldehyde-related cancers among long-term one-pack-per-day tobacco cigarette smokers. If the data are solid, it would be important to acknowledge that increased risk. At the same time, you should offer some perspective by pointing out the dramatic *decrease* in the theoretical risk of many other forms of cancer if a tobacco smoker were to switch to vaping. Without that credibility (and integrity), such an attempt at putting the results into a larger perspective would likely fall on deaf ears.
- **Consider providing media training resources for researchers, both inside and outside industry.** Surveys of scientists, and my own experience providing media training to academics, suggest that many researchers want to explain their research to the public but face barriers such as finding journalists unpredictable (25) and not understanding their needs and motives; fears about how they will look to their peers; and lack of structured practice in explaining their work in simple, accessible terms with helpful analogies (26). A website available to all scientists is one possibility. An

open web-based resource is immune to concerns or potential discomfort about industry support.

- **Go beyond promoting comments from industry sources.** This goes against common wisdom. Industry groups routinely promote their executives or spokespeople to comment on scientific research. However, as shown above, journalists are naturally wary of industry sources (27), and particularly those linked with the tobacco industry.
  One alternative is building a database of highly qualified university-based and independent (non-industry) researchers who are approaching vaping, nicotine and related issues such as youth risk behaviors from the perspective of harm reduction, and who can be referred to journalists to help them interpret research accurately in their areas of expertise. Researchers would of course need to give permission for such referrals, with the shared goal of broadening journalists' access to experts who can improve the public's understanding of science.
- **Develop a list of key journalists** who are likely to cover health and science stories in a variety of media, and build relationships with them. Previous research (28) and the examples above clearly show the difference in the quality of reporting of complex scientific information between those reporters who covered health and science as a beat (and likely had academic training in science or medicine) and those who were general assignment reporters who were asked by their editors to cover a vaping-related science research story. Do not neglect online-only journalists, who may have considerable expertise in health and science (29).
  Because of the significant turnover among media outlets, this list will have to be continually updated. Build relationships with these journalists ahead of time so that they know you can refer them to independent and useful resources (interview subjects) when they are covering a vaping-related story. Let the journalists know that the researchers you'll refer them to are not connected to or receiving funds from the e-cigarette industry, but simply want to promote the accurate coverage of this complex area of science.
  Also remember that there are multiple potential "touch points" (both times and people) for reaching out to news media to increase the odds of accurate interpretations of scientific studies.
- **Develop simple online materials that can help journalists** ask the types of questions about such studies that scientists would ask in order to assess the quality of the data and of the conclusions. There is a need and demand for such training (30), but journalists frequently lack time and money to attend formal training courses.
  In my earlier work at the Harvard Medical School/Massachusetts General Hospital Center for Mental Health and Media, and under a grant from the National Institutes of Health, we developed a training program in

conjunction with the National Press Foundation. It was aimed at helping general assignment reporters in all media improve their critical thinking skills when screening press releases and reporting science-related stories. It was called "Spot the Crap!" (31) That experience taught us that reporters, producers and editors strongly preferred such professional education to be online and available when they need it rather than live classroom instruction.

Multimedia web content might cover topics such as how to communicate about relative risk, a quick-and-dirty guide to statistics, and how to "spot the crap" in misleading press releases about research findings. Materials could include interviews with reporters specializing in health and science who share their reporting tips and give examples of successes (*e.g.*, seeing the connection between a technical research paper and a community issue) and failures (*e.g.*, not being appropriately skeptical of a study's results).

- **Build an online repository of resources** on nicotine, reduced-harm products, tobacco smoking cessation and health that can be accessed by reporters when they need it. These should be from independent (third-party) sources such as research publications and white papers by independent researchers. Thus, for example, when it would be helpful for a journalist to have an authoritative source cite the theoretical (and eventually, clinical) decline in cancer risk when a tobacco smoker switches to vaping, you can quickly provide it.

  Finally, to help non-specialist reporters understand the issues, consider supporting the creation of plain-language summaries of tobacco-related technical reports that are issued to members of the media. For example, media coverage of the 2016 Cochrane Collaboration report, "Electronic Cigarettes for Smoking Cessation," (32) drew heavily on and was greatly improved by the plain-language summary included in their otherwise daunting report.

References

1. National Science Board, National Science Foundation. *Science and Engineering Indicators 2016* (NSB-2016-1), 2016 Jan. https://www.nsf.gov/statistics/2016/nsb20161/#/report/chapter-7/interest-information-sources-and-involvement
2. Yates K., Friedman K, Slater MD, Berman M, Paskett ED, Ferketich AK. A content analysis of electronic cigarette portrayal in newspapers. *Tobacco Regulatory Science*, 2015 Apr., 1(1), 94-102.
3. Patterson C, Hilton S, Weishaar H. Who thinks what about e-cigarette regulation? A content analysis of UK newspapers. *Addiction*, 2016 July, 111(7), 1267-1274.

4. Jensen RP, Luo W, Pankow JF, Strongin RM, Peyton DH. Hidden formaldehyde in e-cigarette aerosols [letter]. *New England Journal of Medicine*, 2015 Jan. 22, 372(4), 392-394.
5. National Toxicology Program Report on Carcinogens, Thirteenth Edition. 2014 Oct. Department of Health and Human Services, Public Health Service, National Toxicology Program.
6. http://www.latimes.com/science/sciencenow/la-sci-sn-electronic-cigarette-formaldehyde-20150121-story.html
7. https://www.wsj.com/articles/study-links-e-cigarettes-to-formaldehyde-cancer-risk-1421877617
8. http://www.npr.org/sections/health-shots/2015/01/21/378663944/e-cigarettes-can-churn-out-high-levels-of-formaldehyde
9. http://www.cbsnews.com/news/e-cigarette-vapor-filled-with-cancer-causing-chemicals/
10. http://www.nbcnews.com/health/cancer/you-vape-high-levels-formaldehyde-hidden-e-cigs-n290826
11. http://www.cbsnews.com/videos/hidden-carcinogen-found-in-e-cigarettes-study-finds
12. https://www.pdx.edu/news/node/27413 and https://www.pdx.edu/clas/e-cigarette-research-faq
13. American Council on Science and Health. "E-cigarette Study Gets Big Headlines, But What Does It Mean?" http://www.acsh.org/news/2015/01/22/poorly-done-e-cig-vapor-study-gets-big-headlines-means-nothing
14. http://www.npr.org/sections/health-shots/2015/01/21/378663944/e-cigarettes-can-churn-out-high-levels-of-formaldehyde
15. http://www.reuters.com/article/usa-health-ecigarettes-idUSL1N0UZ24620150121
16. Dai H, Hao J. Flavored electronic cigarette use and smoking among youth. *Pediatrics*. 2016;138(6):e20162513 http://pediatrics.aappublications.org/content/early/2016/11/03/peds.2016-2513.full
17. http://www.cdc.gov/tobacco/data_statistics/surveys/nyts/
18. http://www.reuters.com/article/us-health-ecigarettes-kids-idUSKBN13223G
19. http://www.webmd.com/smoking-cessation/news/20161107/flavored-e-cigarettes-may-entice-teens-to-smoke-study#1
20. Stimson GV, Thom B, Costall P (editorial). Disruptive innovations: The rise of the electronic cigarette. *International Journal of Drug Policy*, 2014, 25, 653-655.
21. http://www.cdc.gov/media/releases/2015/p0416-e-cigarette-use.html
22. Arrazola RA *et al.* Tobacco use among middle and high school students — United States, 2011–2014. *MMWR*, 2015 April 17, 64(14), 381-385.
23. http://www.capecod.com/newscenter/do-e-cigarettes-help-you-quit-smoking/

24. Steinberg L. A social neuroscience perspective on adolescent risk-taking. *Developmental Review*, 2008 March, 28(1), 78-106.
25. Peters HP. Gap between science and media revisited: Scientists as public communicators. *PNAS*, 2013 Aug. 20, 110 suppl. 3, 14102-14109.
26. Dahlstrom MF. Using narratives and storytelling to communicate science with nonexpert audiences. *PNAS*, 2014 Sept. 16, 111(suppl.4), 13614-13620.
27. Morrell B, Forsyth R, Lipworth W, Kerridge I, Jordens C. Rules of engagement: Journalists' attitudes to industry influence in health news reporting. *Journalism* 2015, 16(5), 596-614.
28. Wilson A, Robertson J, McElduff P, Jones A, Henry D. Does it matter who writes medical news stories? *PLoS Medicine*, 2010 Sept., 7(9), e1000323.
29. Su LY, Akin H, Brossard D, Scheufele DA, Xenos MA. Science news consumption patterns and their implications for public understanding of science. *Journalism & Mass Communication Quarterly*, 2015, 1-20.
30. Voss M. Checking the pulse: Midwestern reporters' opinions on their ability to report health care news. *American Journal of Public Health*, 2002 July, 92(7), 1158-1160.
31. Olson CK, Kutner LA. The public's understanding of research and the role of mass media. *Media Psychology Review*, 2008, 1(1). http://mprcenter.org/review/olson-electronic-media-ed/
32. Hartmann-Boyce J, McRobbie H, Bullen C, Begh R, Stead LF, Hajek P. Electronic cigarettes for smoking cessation. *Cochrane Database of Systematic Reviews*, 2016, 9, Art. No.: CD010216. DOI: 10.1002/14651858.CD010216.pub3

VAPERS AND THE VAPING INDUSTRY ARE THE PRIMARY AGENTS OF TOBACCO HARM REDUCTION IN THE UNITED STATES

Christopher Russell
Centre for Substance Use Research
Glasgow, UK

Abstract

The most commonly used method of quitting smoking in the United States, substituting e-cigarettes for cigarettes, was not conceived by, is not recommended by, and was, until very recently, not controlled by the U.S. Food and Drug Administration (FDA). The increasing popularity of a method of quitting smoking that is not recognised by FDA as an effective smoking cessation method highlights a divergence between how tobacco harm reduction has been conceptualised within the medical science and health policy communities and how tobacco harm reduction is being exercised in the real world by more and more smokers.

In this article, I argue that, for the past decade, tobacco harm reduction in the real world has been driven by manufacturers, vendors, consumers and advocates of vapour products. Testimony reveals that smokers rarely initiate vaping without first researching the products or talking to a vaper or a vape shop sales assistant about why they vape, how the vaping experience compares to smoking, or how their life and health has changed since becoming a former-smoking vaper. The knowledge, practical advice and encouragement that have been transferred to smokers through millions of such interactions have likely rationalised and motivated millions of attempts to substitute e-cigarettes for conventional cigarettes that may not have occurred in the absence of these peer interactions.

The increasing preference of U.S. smokers to use nicotine-containing products as a method of quitting smoking suggests progress towards a smoke-free society may be further accelerated by a nicotine regulatory system that encourages an expansion and diversification of the market in new tobacco and nicotine products that appeal to and reduce harm to smokers, and maximises the user community's opportunities to interact with smokers.

Introduction

Tobacco smoking continues to kill more people, cause more disease, and contribute more to social inequalities in high-income countries than any other preventable factor (1). In the United States, 480,000 lives are lost annually to smoking-related diseases, with more than 20 times this number of people suffering with a smoking-related disease (2,3). Smokers lose around three months of life, on average, for every year of smoking after age 35, equating to around 10 years of life lost by lifelong smokers compared to non-smokers (4,5). Quitting tobacco smoking at the soonest opportunity is therefore the best action a person can take to improve his or her health in the medium to long-term (6), but quitting smoking is notoriously difficult. Given that most of the 8 million smoking-related deaths that are projected to occur globally by 2030 will be among people who are currently smoking, not those who have yet to start (7), developing new and evermore effective ways of helping more people to quit smoking as soon as possible is a public health imperative.

Tobacco harm reduction aims to prevent or reduce harm by promoting substitution of combustible tobacco with less hazardous non-combustible sources of nicotine to smokers who do not quit smoking in response to conventional tobacco control measures (1). E-cigarettes – hand-held devices that use battery power to heat a solution of propylene glycol, glycerol and often flavorings and nicotine, to deliver nicotine to the user via an inhaled aerosol – have rapidly grown in popularity among adults as an alternative to smoking conventional cigarettes. Though data on the safety of long-term use of nicotine by inhalation will not be available until e-cigarettes have been in widespread use for several decades, two national health authorities in the United Kingdom – Public Health England (8) and the Royal College of Physicians (1) – estimate the long-term health risks of inhaling the known constituents of e-cigarette vapor to be unlikely to exceed 5% of the health risks of smoking tobacco. These authorities also agree that e-cigarettes may be amongst the most effective methods for helping people stop or reduce smoking (9,10,11,12). While continuing to recognize complete cessation of all tobacco and nicotine use as the best course of action a smoker can take to improve his or her health, the U.K. Department of Health has therefore pledged to support policies that maximize the availability of less harmful forms of nicotine delivery, such as e-cigarettes, to adult smokers, and support smokers to switch to less harmful nicotine products at the soonest opportunity (13). Given that substitution of e-cigarettes for tobacco smoking has the potential to prevent almost all the harm caused by smoking, identifying initiatives, settings and people that may increase smokers' capability, opportunity and motivation to use e-cigarettes in place of tobacco cigarettes is vital.

Increasing use of e-cigarettes is associated with an increased rate of smoking cessation

E-cigarettes are now the most popular assisted method of quitting smoking in the United States, used in 35% of most recent quit attempts (14). Moreover, a recent analysis of U.S. population survey data for 2014-2015 (15) concluded that e-cigarette users were more likely than non-users to attempt to quit smoking (65% vs. 40.1%) and more likely to succeed in quitting (8.2% v. 4.8%). Data also revealed a statistically significant association between an increase in e-cigarette use between 2010 and 2015 and an increase in smoking cessation at the population level.

Specifically, the population rate of smoking cessation in the United States was 1.1% higher in 2015 – when the population rate of current e-cigarette use was 28.0% – than in 2010 – when the population rate of current e-cigarette use was around 1% – and higher than all previous survey years – when the population rate of current e-cigarette use was non-existent. This 1.1% increase in the smoking cessation rate between 2010 (4.5%) and 2015 (5.6%) is significant for several reasons. First, 1.1% represents approximately 350,000 additional ex-smokers in 2014-2015. Second, such an effect on smoking cessation at the population level has been predicted, but never observed, for nicotine replacement therapies or varenicline (Chantix). Third, 1.1% represents the first significant increase in the smoking cessation rate at the population level in the U.S. for the past 25 years. The study authors caution, however, against attributing the full 1.1% increase to e-cigarettes alone, and instead suggest it is more probable that e-cigarettes interacted synergistically with media campaigns, tobacco tax increases and other state and local tobacco control measures to yield this effect. For example, mass media campaigns may have increased smokers' motivation to quit, while e-cigarettes may have increased the probability of a smoker making an attempt to quit and succeeding.

The apparent positive effect of e-cigarettes on population smoking cessation rate in the U.S. is consistent with that revealed by a 2016 time-series analysis of population trends in e-cigarette use and smoking in England (16). E-cigarette use as part of a quit attempt rose from negligible use in the last quarter of 2006 to 35.0% in the first quarter of 2015. Over the same period, there was an overall increase in the success rate of those who reported a quit attempt (from 10.6% in the last quarter of 2006 to 18.6% in the first quarter of 2015). In all, e-cigarettes were estimated to have contributed approximately 18,000 additional long-term ex-smokers in England in 2015. If the link between change in e-cigarette use and smoking cessation rate was causal, then every 10% increase in use of an e-cigarette as part of a quit attempt would result in a 0.58% increase in successful quit attempts, other things being equal.

Overall, the results of these two studies, based on large representative samples of e-cigarette users in the U.S. and U.K., suggest a similar positive effect on smoking cessation at the population level may be seen in other jurisdictions if a large proportion of smokers can be encouraged and assisted to use e-cigarettes on a daily basis to aid attempts to quit smoking. That is, encouraging and assisting as many smokers as possible to switch to e-cigarette use at the soonest opportunity appears to offer governments the best opportunity in a generation to significantly accelerate the rate at which people quit smoking for good. To the extent that the U.S. Food and Drug Administration (FDA) may ultimately follow the U.K Department of Health in concluding that e-cigarettes are likely to be substantially less harmful to the user and bystanders than conventional cigarettes, and that promoting substitution of e-cigarettes for conventional cigarettes among existing smokers would likely yield a substantial net public health benefit, the critical question for U.S. tobacco control policy then is: who and what has the greatest potential to increase the proportion of U.S. adult smokers who use e-cigarettes as an aid to quitting smoking? To answer this, we must first be clear about who and what has been primarily responsible for growth in the rate of current use of e-cigarettes among U.S. adult current smokers from 1% in 2010 to 30.1% in 2015.

Explaining the rise of vaping in the United States

In this article, I argue that, for the past decade, tobacco harm reduction in the United States has been driven by manufacturers, vendors, consumers and advocates of vapor products. The increasing use of e-cigarettes as aids to quitting smoking since 2010, and the associated increase in smoking cessation at the population level, I suggest, is almost entirely attributable to the availability of increasingly satisfying vapor products, and to the education, advice and support that has been transferred from vapers to smokers through millions of interactions in vape shops, on vaping-dedicated websites, in online discussion groups, at workplaces and at social events. Lastly, I argue that encouraging smokers to engage with and learn from vapers in places that are dedicated to vaping-related discussion, education and advice – in particular, vaping-dedicated websites and vape shops – represents the best opportunity to accelerate the population rate of switching from smoking to vaping in jurisdictions where vapor products are accessible, affordable and attractive to adult smokers.

U.S. public health efforts to discourage e-cigarette use by smokers

Federal, state and local U.S. public health authorities, in general, can claim no credit for the increasing use of e-cigarettes by adult smokers or, by implication, for the significant increase in smoking cessation at the population level that is attributable to an increased prevalence and frequency of use of e-cigarettes to aid smoking quit attempts. Since the emergence of e-cigarettes in the U.S. market, the FDA, Centers for Disease Control and Prevention (CDC), and a long list of medical institutions and health charities (*e.g.* American Lung Association, American Heart Association, Campaign for Tobacco-Free Kids) have explicitly discouraged smokers from using e-cigarettes either as a substitute for conventional cigarettes, an alternative to using nicotine replacement products and medications that are FDA-approved for smoking cessation, or as an alternative to quitting all tobacco and nicotine use completely. The messages consistently communicated to smokers, implicitly and explicitly, by these authorities have been that the long-term risks of using e-cigarettes are not known, e-cigarette vapor contains many of the same chemicals found in cigarette smoke, the relative harmfulness of e-cigarettes and cigarettes has not been established, and the effectiveness of e-cigarettes for helping people to stop smoking has not been proven. It is feasible that repeated exposure to such claims have maintained smoking by individuals who would have considered switching to e-cigarettes, any may be one of several factors that explain why the proportion of U.S. adult current smokers and former smokers who incorrectly believe that using an e-cigarette is equally or more harmful than smoking a conventional tobacco cigarette has increased annually between 2012 and 2106 (17). Given that smokers' decisions to switch to e-cigarettes are known to be strongly influenced by a belief that e-cigarettes are likely to be less harmful than conventional cigarettes, this increasingly common misperception of the harmfulness of e-cigarettes relative to conventional cigarettes may in part explain why around six in ten U.S. adult current smokers have never used an e-cigarette (18).

What has occurred since 2010, therefore, is that millions of U.S. adults have made a decision to quit smoking by using vapor products that are not recognized by any local, state or federal health authority as effective for smoking cessation. The increasing popularity of a method of quitting smoking that, to date, has not been recognized by any health authority in the United States as an effective or potentially effective smoking cessation method highlights a divergence between how tobacco harm reduction has been conceptualized within the medical science and health policy communities and how tobacco harm reduction is being exercised in the real world by more and more smokers.

Vapers and the vape industry are the primary agents of tobacco harm reduction

With the exception of the publication of scientific articles by a small number of tobacco harm reduction scientists that received very limited coverage by news media, the vape-using community and the vape-manufacturing industry have provided the only sources of encouragement for U.S. adult smokers to switch to e-cigarettes in the past ten years. The rapid increase in use of e-cigarettes in the United States is best described as a consumer-led revolution in how nicotine is consumed in society. This assertion is true in two senses: first, vapor products emerged as a widespread alternative to cigarettes in response to smokers' want for inhalational products that decoupled low-risk nicotine from high-risk smoke, and second, many individuals became not merely consumers of vapor products, but enthusiasts and advocate who took it upon themselves to encourage and assist more smokers to try vaping.

The modern history of vaping began in the early 2000s when a Chinese pharmacist, Hon Lik, motivated by the death of his father to lung cancer caused by smoking, conceived and invented the electronic cigarette as a means to accelerate the decline in cigarette smoking, and more immediately, by aerosolizing nicotine for inhalation via vapor rather than smoke, to reduce and prevent harm to tobacco smokers. Lik believed that if e-cigarettes could be manufactured to deliver nicotine with a speed and efficiency that rivalled nicotine delivery via tobacco smoke, and these devices were made available, attractive and affordable to all smokers in the population, many smokers would choose to use an e-cigarette in place of conventional cigarettes. And so, with the help of his employer, Ruyan Group, now Dragonite International, Lik brought the e-cigarette first to Europe, and then in 2007, to the United States. Early e-cigarette models quickly became popular with smokers who wanted to consume nicotine in a way that replicated the experiential satisfaction, sensorial pleasures, behavioral rituals, and emotional functions of smoking a cigarette, that, critically, wouldn't kill them. This new technology gave smokers hope that further innovations of vapor technology could eventually yield a portfolio of vaping products that deliver 'all of the pleasure of smoking with none of the death'. Since Lik's early e-cigarette, manufacturers have aspired to innovate nicotine vaporizing technologies ever closer to meeting this pleasure-health ideal.

As vape manufacturers have innovated, expanded and diversified a portfolio of vaping products to better meet the wants and needs of more smokers, and as users have reaped subjective health, social and financial benefits from the use of these products, online and offline communities comprising millions of former-smoking vaping enthusiasts from around the world have formed. Beyond discussion of new devices and flavors among advanced vapers, these communities are typically characterized by a selfless desire, even a perceived responsibility, to educate, advise and support smokers and new vapers to get started with vaping and to get the most out of vaping, such that giving up smoking becomes experientially easier as vaping continues.

The remainder of this article will make the case for why vapers, the actual users of vapor products, are best placed to increase e-cigarette use by smokers, and suggest where vapers may most effectively encourage, educate, and assist smokers to become vapers.

Encourage smokers to engage with vapers online

Testimony reveals that smokers rarely initiate vaping without first researching the products or talking to a vaper or a vape shop sales assistant about why they vape, how the vaping experience compares to smoking, or how their life and health has changed since becoming a former-smoking vaper. The learning, practical advice and encouragement that has been transferred to smokers through millions of such interactions with vaping peers have likely been critically important in rationalising millions of smokers' decisions to substitute e-cigarettes for conventional cigarettes, and motivating quit attempts that would not have occurred without these peer interactions. Even as interest in vaping increases among medical practitioners and public health scientists, the vape-using community will continue to be uniquely, and perhaps, best qualified to advise and educate smokers, researchers and regulators about the true nature and character of vaping.

Given experienced vapers' apparent eagerness to share with smokers the types of advice and information that answer smokers' most frequently asked questions, the potential of former-smoking experienced vapers to assist smokers to switch to e-cigarettes may be substantial. The question then is, in which settings, and through which media, would smokers be most interested to interact with vaping peers about vaping, and vapers be most accessible to smokers? A first reaction among public health researchers may be to suggest that primary health services and smoking cessation services are the optimal settings in which vapers can intervene to help smokers switch. Though vapers may have significant potential to boost quit rates within smoking cessation services by helping smokers in ways that smoking cessation advisors cannot – and longitudinal studies of this question are warranted – the suggestion that vapers' potential to convert smokers would be best realized within formal health service settings fails to recognize that millions of smokers have become vapers over the past decade without ever interacting with the public health community on vaping, or even in spite of their doctors' warnings against the use of e-cigarettes.

Instead, the rapid growth in the number of ex-smokers in the U.S. who now vape has occurred through millions of interactions between smokers and the vape community – manufacturers, vendors and consumers of vapor products. These peer-to-peer interactions take place primarily where vapor products are sold (*e.g.* vape shops) and online at websites dedicated to discussion and learning about vaping (*e.g.* discussion fora, blogs, instructional articles, tutorial videos). The authenticity, specificity and accessibility of the education about vaping available through interactions with vapers in vape shops and vaping-dedicated websites are likely among the main reasons why smokers and new vapers may tend to prefer to learn about vaping from the vape community than from the public health community. And in the past decade, thousands of free online resources have been created by experienced vapers to provide education, practical advice and motivational support to smokers who are curious about vaping or contemplating a switch to vaping. The remainder of this article will describe how three vaping-dedicated websites have dwarfed every public health resource about vaping in terms of their reach, attractiveness, availability, education and advice on getting started with vaping and switching completely from smoking to vaping.

E-Cigarette Forum

E-Cigarette Forum (ECF; www.e-cigarette-forum.com) has become the largest e-cigarette specific social network since its founding in 2007, with now more than 254,000 members, of whom approximately 10% are active at a given time, and with approximately 1.9 million unique visitors per month. Together, these members and guests have posted more than 17.7 million messages to more than 632,000 separate discussions about vaping, most of which are written by advanced vapers for the benefit of smokers and new vapers. Smokers may wish to start by reading messages posted in the larger discussion threads (*e.g. 'General Vaping Discussion'* or *'General E-Liquid Discussion'*) or in discussion threads more relevant to their specific questions and concerns (*e.g. 'Vaping Success Stories', 'Health, Safety and Vaping', 'Requests for Opinions/Reviews'*, or *'Ask the Veterans'*). At any time of the day, without leaving their house or making an appointment, smokers can learn how to get started in using thousands of specific brands and models of vaping hardware and e-liquids, and initiate conversations with experienced actual users of specific devices, kits and liquids. A true public health benefit of the discussion threads of ECF is that no smoker in the world need every try a vapor product without first being able to talk to tens if not hundreds and thousands of people who have used that product, and are willing to give their review of the product. Indeed, one of the main points of advice given by members of ECF to new vapers is to 'ask before you buy'. Such preliminary advice can in many cases save a smoker from purchasing a poor-quality product or a product that just won't, for example, give him/her the flavor experience or volume and vapor he/she is looking for.

Vaping360

Vaping360 (www.vaping360.com) was established in August 2015 as a platform for smokers, new vapers, researchers, nicotine regulators and other stakeholders to learn from vapers about the best (and worst) ways to transition from regular smoking to regular vaping. The site has since evolved into a more holistic source of education and assistance for smokers trying to switch. Vaping360 now features regularly posted plain English summaries of new published scientific research on the health effects of vaping; plain English summaries of legislation and regulations that may affect the availability, affordability and attractiveness of vapor products; plain English reviews of newly released vaping devices and flavors; news of vaping advocacy events taking place in various cities and states; and a synthesis of live deals and discounts on vapor products from across the internet.

A main feature of Vaping360 is called 'Best Vapes', which, as the title suggests, rates the 'best' three products in a range of vapor product categories, from the *'Best Starter Kits'* for smokers to switch to vaping and the *'best e-juice brands to try on a budget*, to slightly more advance products, such as *'Best Vape Pens'* and *'Best Clearomizers'*, to yet more advanced vapor products, such as 'Best Rebuildable Tank Atomizers' and 'Best Sub-Ohm Tanks'. The status of the three products in each category as the 'best' is based on reviewers' ratings of each product on a number of criteria, such as flavor, air flow, clouds, battery life and value for money. Such recommendations can help smokers, new vapers, and advanced vapers 'see the wood for the trees' by more quickly narrowing down choices to products that are more likely to be 'right' for them.

Vaping360 also features a 'How to' section for smokers who are contemplating switching to vaping and new vapers, called *'Vaping 101'*. This section contains short articles, written by advanced vapers, giving practical advice and answers to some common questions about how to get started with vaping. Recent articles include

'10 simple steps to stop your tank leaking', '*PG (propylene glycol) vs VG (vegetable glycerin)*', '*Rebuildable tanks explained*', and '*Vaping and Inhaling: Everything you need to know*', among many others. These articles are contributed by more than 20 freelance writers from around the world, almost all of whom are advanced vapers who used to smoke.

Like ECF, what makes Vaping360 a resource capable of educating and encouraging smokers at the population level to switch to vaping is its huge global popularity with vapers and potential for attracting smokers in the U.S. and around the world – the site has been visited by approximately 15.5 million unique individuals from around the world since its launch in 2015, and is visited by approximately 1.1 million unique individuals every month, of whom approximately 500,000 come from the United States.

Vaping.com

Vaping.com (VDC), a sister site to ECF, is best known to vapers as an online 'supermarket' for purchasing popular brands of vaping kits, devices, liquids, tanks and parts. But VDC has also become known as the home of an excellent series of 'Vaping Guides' targeted at smokers and new vapers. These 'Guides' are regularly posted short articles, written by experienced vapers in simple non-technical language, that give practical step-by-step advice, recommendations, and answers to some of the most fundamental questions asked by new vapers, and to some questions that new vapers may not think to ask. Recent articles include '*Mouth to Lung vs. Direct to Lung. All about puff; which is for you?*' – in which the author explains the differences between two common puffing styles, and the benefits and sensations associated with each – '*How to Clean a Vape Device: The Newbie Maintenance Guide*' – which gives detailed instructions on how to simple rinse, deep clean and re-assemble the parts of a vape tank – '*The Biggest Mistake New Vapers Make*' – which, according to the author, is failing to carry a backup device or failing to carry additional supplies to service one's main device – and '*Start Vaping in 5 Easy Steps, the Beginners Guide*'. For those interested, the author of the latter article guides smokers on how to pick the experience they are looking for, how to pick a device that will best meet their wants and needs, how to pick flavors and nicotine strengths they will enjoy vaping, how to get the right parts to maintain their chosen device (compatible replacement coils and batteries), and to read the instructions that come with their chosen products to ensure they get the most satisfaction out of using the products.

Reading these four short but incisive articles on VDC would take a smoker but 20 minutes, after which he/she may be expected to be significantly more knowledgeable and confident to choose, use, maintain and enjoy a vaping device. Encouraging smokers to read these Guides, and more widely interact with and seek advice from experienced vapers at these and similar websites, could have significant potential to increase smokers' awareness of the variety of vaping products and features, increase smokers' motivation to try using e-cigarettes in place of cigarettes, and ultimately, help more smokers to find the combination of vaping products that 'work for them' as an alternative to smoking.

Recommended reading for healthcare professionals who interact with smokers

The discussion threads and educational guides to vaping posted at ECF, Vaping360 and Vaping.com are not only informative to smokers and new vapers; they would make excellent recommended reading for smoking cessation advisors and primary care health professionals who regularly interact with smokers but do not feel confident or competent to discuss vaping products with them. Such online information can educate health professionals about, among others, vaping vernacular, the evolution of vaping products, and changing trends in their clients' patterns and styles of vaping. The staff of smoking cessation services and other primary care services may also seek to learn from, and even employ in a voluntary or paid capacity, local former-smoking e-cigarette users and local vendors of vaping products. Participants qualitative responses in this study suggest they may not only be keen to help smokers directly, but would also possibly welcome opportunities and invitations from smoking cessation and other health services to educate staff about vaping, even if in an informal, infrequent, unpaid capacity. At present, both consumers and vendors of vaping products remain resources largely unused by most stop smoking services (SSS) and other primary health services, which not only leaves staff without a valuable source of education, but may also reduce a smoker's interest in engaging with the service. To dismiss or neglect the insights and experiences of these individuals could prove to be a significant missed opportunity for health services, academics and e-cigarette manufacturers to help more people quit smoking and avoid relapse. Longitudinal studies are warranted to examine the extent to which the receipt of support from vaping peer, whether brief or ongoing and whether delivered in person or online, can increase smokers' odds for quitting smoking or cutting daily cigarette consumption, particularly among smokers who have tried and failed to quit numerous times with the use of approved NRTs and smoking cessation medications.

Conclusions

Substitution of low-risk nicotine products, like e-cigarettes, for cigarette smoking is now the most common method of quitting smoking in the United Kingdom and in the United States. Former-smoking vapers, working in vape shops, engaging online in discussion groups, and taking time to offer education, practical advice and tales of personal experience about vaping to friends and colleagues who smoke, have, through their promotion of e-cigarettes as an alternative to smoking cigarettes, appear to be the primary agents of the most successful movement to reduce smoking-related harm in the United States in a generation.

By sharing their stories of how they used an e-cigarette to quit smoking, what they experienced along the way, and how their lives have changed since they switched, former-smoking vapers are providing current smokers with a valuable source of encouragement, knowledge, advice, and motivation required to trial an e-cigarette and then transition fully from smoking to vaping. Indeed, it may not be unreasonable to suggest that connecting a former-smoking e-cigarette user with a smoker who is curious about using an e-cigarette for smoking cessation may turn one quitter into two (*i.e.* 'quitters creating quitters'). Encouraging smokers and new vapers to interact with experienced e-cigarette users who have already quit smoking by switching to e-cigarettes in settings that appeal to them – whether in vape shops, vaping-dedicated websites or within smoking cessation services – may represent an opportunity to accelerate the current rate of switching from cigarette smoking to e-cigarette use.

There continues to be a pressing need for research that collects, summarizes and communicates the views and experiences of successful former-smokers to both current smokers and to those charged with regulating nicotine vapor products. Stories of how an individual switched from smoking to vaping can be hugely informative to smokers and nicotine regulators about the ways in which e-cigarettes have enabled people to quit smoking. If the innovation of increasingly effective and safer e-cigarette products is not stifled by regulations in a jurisdiction, and if smokers are encouraged and supported to use e-cigarettes as an alternative to smoking, educated about the relative health risks and benefits of switching to vaping versus continuing to smoke, and educated on how to properly use and maintain their device, then the individual and population health benefits of e-cigarettes have the potential to be unmatched by any other currently available smoking cessation product, method or policy. Connecting smokers to the insights and advice of former-smoking vapers may represent a significant portion of that potential.

References

1. Royal College of Physicians. *Nicotine without smoke: Tobacco harm reduction*. London: RCP, 2016
2. U.S. Public Health Service, Office of the Surgeon General, National Center for Chronic Disease Prevention and Health Promotion, Office on Smoking and Health. The health consequences of smoking—50 years of progress: a report of the Surgeon General. Atlanta, GA: DHHS, CDC, National Center for Chronic Disease Prevention and Health Promotion, Office on Smoking and Health; 2014.
3. Rostron BL, Chang CM, Pechacek TF. Estimation of cigarette smoking-attributable morbidity in the United States. *JAMA Intern Med* 2014;174(12):1922–1928. http://dx.doi.org/10.1001/jamainternmed.2014.5219. Accessed 21 January 2017.

4. Doll R., Peto R., Boreham J., Sutherland I. Mortality in relation to smoking: 50 years' observations on male British doctors. *BMJ*, 2004; 328:1519–27.
5. Pirie, K., Peto, R., Reeves, G.K., Green, J., Beral, V. The 21st century hazards of smoking and benefits of stopping: a prospective study of one million women in the UK. *Lancet*, 2013; 381: 133-141.
6. Fiore, M. C., Jaén, C. R., Baker, T. B., *et al. Treating Tobacco Use and Dependence: 2008 Update*. Clinical Practice Guideline. Rockville, MD: U.S. Department of Health and Human Services. Public Health Service. 2008.
7. Eriksen M, Mackay J, Ross H. *The Tobacco Atlas. Fourth Edition*. American Cancer Society and World Lung Foundation. 2012.
8. Public Health England. *E-cigarettes: an evidence update*. 2015. Accessed 10 February 2016 at: https://www.gov.uk/government/uploads/system/uploads/attachment_data/file/457102/Ecigarettes_an_evidence_update_A_report_commissioned_by_Public_Health_England_FINAL.pdf
9. McRobbie, H., *et al. Electronic cigarettes for smoking cessation and reduction*. Cochrane Database Systematic Review, 2014; 12: CD010216.
10. Brown, J., Beard, E., Kotz, D., Michie, S., West, R. Real-world effectiveness of e-cigarettes when used to aid smoking cessation: A cross-sectional population study. *Addiction*, 2014; 109:1531-1540.
11. Beard, E., West, R. Michie, S. Brown, J. Association between electronic cigarette use and changes in quit attempts, success of quit attempts, use of smoking cessation pharmacotherapy, and use of stop smoking services in England: time series analysis of population trends. *BMJ*, 2016; 354:i4645.
12. Farsalinos, K., Poulas, K., Voudris, V., Houezec, J. Electronic cigarette use in the European Union: Analysis of a representative sample of 27460 Europeans from 28 countries *Addiction*, 2016; 111(11): 2032-2040.
13. U.K. Department of Health. *Towards a smokefree generation - A tobacco control plan for England*. July 2017. Accessed 10 July 2017 at: https://www.gov.uk/government/publications/towards-a-smoke-free-generation-tobacco-control-plan-for-england
14. Caraballo RS, Shafer PR, Patel D, Davis KC, McAfee TA. Quit Methods Used by US Adult Cigarette Smokers, 2014–2016. *Prev Chronic Dis* 2017;14:160600. DOI: http://dx.doi.org/10.5888/pcd14.160600
15. Zhu, SH., Zhuang, YL., Wong, S., Cummins, S.E. Tedeschi, G.J. E-cigarette use and associated changes in population smoking cessation: evidence from US current population surveys. *BMJ* 2017; 358:j3262.
16. Beard, E., West, R., Michie, S., Brown, J. Association between electronic cigarette use and changes in quit attempts, success of quit attempts, use of smoking cessation pharmacotherapy, and use of stop smoking services in England: time series analysis of population trends. *BMJ* 2016; 354:i4645.
17. Majeed, B.A., Weaver, S.R., Gregory, K.R. *et al.* Changing Perceptions of Harm of E-Cigarettes Among U.S. Adults, 2012–2015. *Am J Prev Med* 2017; 52(3); 331-338.

18. Coleman, B.N., Rostron B., Johnson, S.E. *et al.* Electronic cigarette use among US adults in the Population Assessment of Tobacco and Health (PATH) Study, 2013–2014. *Tob Control*, 2017, Jun 17, pii: tobaccocontrol-2016-053462. doi: 10.1136/tobaccocontrol-2016-053462. [Epub ahead of print]

Note that this conference had to be rescheduled due to a hurricane. As a result, several papers were withdrawn, and several had to be presented by the author's colleague.

5. USE OF FORMALIN IN E-VAPOR PRODUCTS TO MONITOR FORMALDEHYDE DELIVERY. Kathleen SPANGLER, James Wilkinson, Matt Melvin, John H. Miller IV and Georgios D. Karles; Altria Client Services, Richmond, VA, USA

Guidance provided by the FDA (May of 2016) for Premarket Tobacco Product Applications for Electronic Nicotine Delivery Systems proposes full chemical characterization of the e-cigarette aerosol. One class of compounds the FDA recommends to be reported in aerosol are the carbonyl containing compounds, formaldehyde, acetaldehyde, acrolein and crotonaldehyde. There currently is no CORESTA recommended method for measuring carbonyls in e-vapor products; however, the CORESTA E-vapor Subgroup is currently working on evaluating multiple methods for the analysis of these carbonyl compounds in e-vapor products (e-liquids and aerosols). In order to evaluate if a new method is suitable for e-vapor products, experiments such as trapping efficiency, recovery, repeatability and reproducibility need to be assessed. Since the levels of carbonyls in e-vapor aerosols are typically very low with variable consistency, evaluation of the fore mentioned method performance parameters present unique challenges. In our lab, we have assessed the addition of formalin to a control e-liquid at known levels to serve as a monitor for formaldehyde analysis in the aerosol of e-vapor products. We have demonstrated the stability of formalin measured in e-vapor aerosol for at least six weeks when refrigerated at 4 °C. Additional work will be presented to reinforce that the addition of formalin to e-vapor monitor cartridges provides a direct way for

71st Tobacco Science Research Conference (TSRC)
28 November - 1 December 2017 / Bonita Bay, FL, USA

laboratories to evaluate method robustness for carbonyl analysis and to monitor laboratory performance over time.

6. CHARACTERIZATION OF AEROSOLS GENERATED FROM A PROTOTYPE E-VAPOR PRODUCT AND A CAPILLARY AEROSOL GENERATOR. Jingjie ZHANG, Christopher Tucker, K. Monica Lee and Michael Oldham; Altria Client Services, Richmond VA, USA

E-Vapor products, which are gaining steady popularity among tobacco consumers, are composed of various types of e-liquid formulations and aerosol generating devices. There are situations that formulations need to be tested separately from devices. The purpose of this study was to characterize the chemical composition and particle size distribution of the aerosols generated from prototype one-channel CAG and a prototype e-vapor product using an identical e-liquid formulation. Prototype e-vapor products were puffed (55 ml volume, 5 second duration, every 30 seconds) using a 20-port linear smoking machine. The CAG temperature was set to 250 °C to match the measured coil temperature during puffing of the prototype e-vapor product. Aerosols from each device were collected on Cambridge filter pads and analyzed (GC/FID; GC/TCD) for major formulation ingredients (propylene glycol [PG], glycerin, water, and nicotine). Analysis (UPLC/MS) of selected carbonyls (formaldehyde, acetaldehyde, acrolein, and crotonaldehyde) was performed by collecting aerosols using a Cambridge filter pad and an impinger in series. Particle size distribution was measured using a cascade impactor. Similar levels of PG, glycerin, and nicotine were detected in the aerosol from both devices. When detected above the limit of quantification, the levels of selected carbonyls were lower in the prototype e-vapor product aerosols. Both devices generated similar particle size distributions. The CAG and the prototype e-vapor product are comparable in terms of aerosol output (chemical composition and particle size distribution). Therefore, the CAG can be used as a surrogate device to generate and test e-vapor formulations in in-vitro and in-vivo studies.

7. NEUTRAL RED UPTAKE CYTOTOXICITY ASSAY IN A549 CELLS UNDER DIFFERENT CULTURE CONDITIONS. Bonnie G. COFFA, Utkarsh Doshi, Jingjie Zhang, Willie J. McKinney, K. Monica Lee; Altria Client Services, Richmond, VA, USA and Pooja Desai; Enthalpy Analytical, Richmond, VA, USA

The OECD guideline (TG129) recommends mouse fibroblasts (3T3) & normal human epidermal keratinocytes (NHK) for performing in vitro cytotoxicity assays but also mentions limitations of these cells, such as having little to no metabolic capacity or possibility of inappropriate interpretation of data if these cells are unrelated to the target organ of interest. Considering the primary target organ for inhalable products, we conducted this study to evaluate the suitability of a human derived alveolar cell line A549 for the neutral red uptake (NRU) assay under various testing conditions. First, positive control sodium lauryl sulfate (SLS) and DMSO extracted cigarette smoke total particulate matter (TPM) were tested for cytotoxicity in A549 and 3T3 under submerged condition. Secondly, A549 were grown at the air-liquid interface (ALI) and assessed for cytotoxicity following 24 and 48 hr. of treatment with SLS in media supplied from basal side, containing 0% or 5% serum. Dose-dependent cytotoxicity was observed in 3T3 and A549 in response to both SLS and TPM and the corresponding IC50 were estimated. Under submerged conditions containing serum, A549 displayed lower IC50 (SLS) and higher IC50 (TPM) in comparison to the values of mouse 3T3 cells, suggesting species-dependent differences in sensitivities. No significant difference in cytotoxicity was observed in A549 cells between submerged and ALI cultures in response to SLS. Enhanced cytotoxicity was evident in A549 cells exposed to SLS under serum free environmental conditions. In summary, 1) A549 cells could be effectively used in the TG129 NRU assay in submerged as well as ALI conditions and 2) media composition such as presence of serum, should be carefully controlled especially when using human-derived cells such as A549.

Note that the 71st TSRC had to be rescheduled due to a hurricane. As a result, several papers were withdrawn, and several had to be presented by the author's colleagu

8. ARE AVAILABLE TEST METHODS FOR THE DETERMINATION OF CARBONYLS IN MAINSTREAM CIGARETTE SMOKE FIT FOR THE ANALYSIS OF CIGARS? Regina M. BALLENTINE, Karen C. Avery, Matt S. Melvin, Jennifer H. Smith and Karl A. Wagner; Altria Client Services, Richmond, VA, USA

In May 2016, the U.S. Food and Drug Administration (FDA) issued a final rule to deem cigars to be subject to the Federal Food, Drug, and Cosmetic Act, as amended by the Family Smoking Prevention and Tobacco Control Act. As part of this regulation, the FDA will require manufacturers to report the quantities of Harmful and Potentially Harmful Constituents (HPHCs) in cigar filler and smoke. Standardized methods do exist for the analysis of carbonyls in cigarette smoke; however, these methods have not been shown to be fit for purpose for the analysis of cigars. CORESTA Recommended Method No. 74, "Determination of Selected Carbonyls in Mainstream Cigarette Smoke by High Performance Liquid Chromatography" was based on Health Canada method T-104 and is the basis of ISO/CD 21160:2017, "Determination of selected carbonyls in the mainstream smoke of cigarettes -- Method using High Performance Liquid Chromatography". Due to the fact that it may take an hour or more to collect cigar smoke, we hypothesized that the carbonyl–DNPH derivative would degrade over time, resulting in decreased carbonyl yields. Cigar smoke was collected and time studies were conducted to determine carbonyl stability in the acidic DNPH trapping solution. The result of this work indicates that there is degradation of the DNPH derivative during smoke collection. Results will be presented for different cigar blend types. Furthermore, these results indicate that specialized methods need to be developed for the robust analysis of carbonyls in cigar.

9. THE CHALLENGES OF MACHINE SMOKING THE DIVERSE CIGAR PRODUCT CATEGORY. Tammy L. Blake[1], Regina M. Ballentine[1], Karen C. Avery[1], Anthony P. BROWN[1], Karl A. Wagner[1], Michelle J. Carpenter[2] and Kerry L. Stutt[2]; [1]Altria Client Services, Richmond, VA, USA; [2]Eurofins Lancaster Laboratories, Richmond, VA, USA

In May 2016, the U.S. Food and Drug Administration (FDA) issued a final rule to deem cigars to be subject to the Federal Food, Drug, and Cosmetic Act (the FD&C Act), as amended by the Family Smoking Prevention and Tobacco Control Act (Tobacco Control Act). As part of this regulation, the FDA will require manufacturers to report the quantities of Harmful and Potentially Harmful Constituents (HPHCs) in cigar filler and smoke. The ability to machine smoke cigars is critical for generating meaningful HPHC data. The CORESTA Cigar Smoking Methods Subgroup has been the main driver for the development of cigar smoking methods. This Subgroup has published several CORESTA Recommended Methods (CRMs) which enable the determination of nicotine-free dry particulate matter (NFDPM), nicotine, water, and carbon monoxide. The diversity in shape and size of the cigar product category creates challenges for smoke collection and often requires product specific solutions to achieve acceptable analytical results. Custom cigar holders that meet the requirements of CRM No. 64 will be described and compared to the commercially available cigar holders. The custom cigar holders enable the collection of smoke from untipped and tipped cigars and provide for improved ease of use and reduced variability over the commercially available holders. A comparison between the two styles of cigar holders for the determination of smoke yields for 9 commercial cigar products will be presented.

10. DIFFERENCES IN PLASMA NICOTINE PHARMACOKINETIC PROFILES FOR VARIOUS E-VAPOR PRODUCTS USED BY ADULT SMOKERS UNDER AD-LIBITUM VS. CONTROLLED USE CONDITIONS. Jianmin LIU, Qiwei Liang, Maria Gogova, Yuxi Zhao and Mohamadi Sarkar; Altria Client Services, Richmond VA, USA

Plasma nicotine pharmacokinetic (PK) profiles are often used to characterize nicotine exposure from e-vapor products (EVP). The purpose of this analysis was to determine whether PK profiles are different when adult smokers (AS) use EVPs under ad libitum or controlled use conditions. We conducted a 6-way randomized crossover study in twenty-four AS smoking $\geq$10 cigarettes/day and had not used EVPs in the past month. AS used six different types of EVPs (tank- or cartridge-based) with different flavors and levels of nicotine under two use conditions (10 hours apart) – controlled use of 10 inhalations of 4-second duration with 60-second intervals (over ~10 minutes) and ad libitum use for 10 minutes. Nicotine plasma levels were measured periodically for 120 minutes. The maximum concentration (Cmax0-2hrs) and area

Note that the 71st TSRC had to be rescheduled due to a hurricane. As a result, several papers were withdrawn, and several had to be presented by the author's colleague.

under the curve (AUC0-2hrs) were higher under ad libitum vs. controlled use. On average, AS took ~twice as many puffs under ad libitum compared to controlled use and average puff duration ranged from 3.0 to 3.6 s for the six EVPs. The inter-individual variability (CV%) for Cmax0-2hrs was larger under ad libitum (53%) than under controlled use (28%). The identical use conditions under controlled use may provide a better method to compare nicotine PK from different types of EVPs. The high inter-individual variability under ad libitum use conditions may reflect behavioral aspects e.g. flavor preference and satisfaction. The pros and cons of each test condition for assessing nicotine PK will be discussed. These variability estimates may be used to design future studies with EVPs.

11. MARKTEN® E-VAPOR STABILITY STUDY: DESIGN AND INITIAL DATA. Susan PLUNKETT, Michael Morton, William Gardner, Karl Wagner and Diana McKinney; Altria Client Services, Richmond, VA, USA

The Food and Drug Administration's (FDA's) Draft Guidance for Industry Premarket Tobacco Product Application for Electronic Nicotine Delivery Systems states that an application should include data informing "the established shelf life of the product and changes in pH and constituents (including HPHCs and other toxic chemicals) over the lifespan of the product." The FDA has not published industry guidance on stability testing protocols for e-vapor products, and no scientific literature has been published to date on the design and execution of e-vapor product stability studies. Although e-vapor products are not pharmaceutical products, in the absence of specific guidance, we have adopted the stability study design from the FDA Guidance for Industry Q1A(R2) Stability Testing of New Drug Substances and Products.

We describe the design and share initial data from an e-vapor stability study on MarkTen® e-vapor products analyzing for 50 constituents in both liquid and aerosol. We determined these constituents in 3 batches of product with 10 replicates each in both liquid and aerosol over multiple time points. Two vaping regimens and 2 storage conditions were tested at select time points.

We will discuss constituent levels, trends, and possible constituent sources. Due to the lack of standardized methods, we suggest best practices for data collection and interpretation including blanks to distinguish background levels from the very low levels of e-vapor constituents. Finally we discuss the statistical utility of 10 replicates.

12. NEW SMOKING MACHINE APPROACH FOR CIGARETTE SMOKING UNDER INTENSE AND ISO 3308 CONDITIONS. (Presented by Whitley Lee) Nils Rose; Borgwaldt KC, Hamburg, Germany

ISO 3308 specifies the definitions and conditions for routine analytical smoking machines, defining the puffing regime as a 35ml bell shape puff with a 2s duration taken every 60s. It also specifies two machine types "A" and "B" dependent to the adjustment of the termination devicebut misleadingly distinguished as "rotary machine" and "linear machine" . These were harmonized regarding their tar and CO deliveries in 1995 and 2002.

In 2007, by demand of the WHO, ISO TC 126 started the preparation of a more intensive smoking regime based on a 55ml bell shape puff with a 2s duration taken every 30s and sealed ventilation holes.

In 2010 ISO TC 126 WG10 carried out a study to measure the tar, nicotine and carbon monoxide yields of cigarettes using both smoking regimes. The outcome of this study has shown significant differences in water and corresponding deliveries (TPM,) with a trend to higher deliveries on linear machines. One possible explanation for this phenomenon was assumed to be the distance between the cigarette and the filter pad causing loss due to pre-condensation. The study also shows lower variation in the data gathered for rotary machines.

In consideration of this, a 4-channel hybrid machine was developed. It is tailored to the requirements of the new smoking method and combines the advantages of both machine types in an ideal and most efficient manner.

Note that the 71st TSRC had to be rescheduled due to a hurricane. As a result, several papers were withdrawn, and several had to be presented by the author's colleagu

This poster briefly describes the design approach of the machine and discusses first validation data for both smoking regimes compared to the yields delivered from standard linear and rotary smoking machines.

13. "TRACK AND TRACE" IN THE SMOKING LABORATORY. (Presented by Whitley Lee) Nils Rose; Borgwaldt KC, Hamburg, Germany

"Track and Trace" is one of the major requirements regarding "Good Laboratory Practice". During the whole production process, from incoming goods up to the packed product, cigarettes are well tracked. But in the smoking laboratory this becomes more problematic. The products have to be linked carefully to complex smoking plans including multiple machines and different parameter sets. And finally the products are converted physically into loaded smoke traps continuing their way to the chemical analysis.

Along the way there are plenty of possibilities to mix up samples or data or to lose the link between the traps and the primary product.

This poster describes a technical solution of consistent tracking during the whole procedure to avoid errors and to gain information about the current status of the product during the whole process, which is an important topic for a simplified laboratory accreditation. It also simplifies the laboratory process and offers not only the possibility for higher efficiency but also for greater process **automation.**

14. ANALYSIS OF ISOTOPICALLY LABELED GLYCEROL TO MEASURE EXPOSURE FROM E-CIGARETTES. (Presented by Erika Pfaumiller) Elizabeth M. DIBBERN, Kafonek, C.J., Brown, G.P., Dzerk, A.M., Nachi, R., Newland, K. E., and R. Islam; Celerion, Lincoln, NE, USA

Electronic cigarettes (ecigs) have become a popular mode of nicotine ingestion. They are often promoted as being less harmful than traditional cigarettes, but research supporting that claim is incomplete. The main chemical components of ecigs are propylene glycol and glycerol and they are usually found in a 70:30 or 50:50 ratio with 0.6 to 3.6% nicotine. Recently, Celerion Inc. was granted a patent for using isotopic propylene glycol to monitor how much PG is ingested from ecig use (vaping). The use of isotopic propylene glycol made clear delineation of ingestion of propylene glycol from foods versus ecigs. A similar method was explored for the quantitation of glycerol. Glycerol has the added challenge that lipids from the plasma – specifically triglycerides - break down into glycerol, greatly increasing the basal glycerol concentrations compared to propylene glycol concentrations. Literature values for free glycerol concentrations range from 6500-12000 ng/mL(2-4) and after breakdown of the triglycerides, total glycerides are more than 10X that. Assuming uptake similar to the propylene glycol, the Cmax for glycerol vaping would be 600-2000 ng/mL. The method detailed here deftly avoids isotopic contribution from endogenous glycerol; the result is an accurate, reproducible method for isotopic glycerol in human plasma.

16. WEB-BASED SYSTEM FOR ORDERING TOBACCO REFERENCE PRODUCTS AND MANAGING PROFICIENCY STUDIES. Bront DAVIS and James Hunsucker; University of Kentucky, Lexington, Kentucky, USA

The Center for Tobacco Reference Products in the College of Agriculture, Food and Environment at the University of Kentucky has developed a website (https://ctrp.uky.edu) to streamline the sale and distribution of tobacco reference products and proficiency test kits. Users must request access to the secure website, and once granted, they are able to purchase products and manage purchases via the website. Significant design considerations and user feedback have been incorporated in the proficiency testing functionality to improve the reporting process for participants, including: (1) creation of the Excel data templates for proficiency studies; (2) purchasing proficiency studies; (3) accessing the Excel data template; (4) uploading Excel data with an opportunity to review before final submission; (5) accessing interim report and commenting on the process; and (6) accessing final reports. This online system has significantly improved the data collection, analysis and reporting process associated with conducting proficiency testing programs.

Note that the 71st TSRC had to be rescheduled due to a hurricane. As a result, several papers were withdrawn, and several had to be presented by the author's colleague.

17. STABILITY OF THE CERTIFIED 1R6F REFERENCE CIGARETTE. Huihua JI, Ying Wu and Franklin Fannin; University of Kentucky, Lexington, Kentucky, USA

University of Kentucky, Center of Tobacco Reference Products (CTRP), has provided reference cigarettes for almost 50 years. These reference cigarettes are widely used as control samples in tobacco research including analytical method development and modified risk tobacco product development. In 2014, CTRP obtained a service agreement with the US Food and Drug Administration (FDA) to produce a certified reference cigarette. The first certified reference cigarette, 1R6F, was manufactured in March, 2015. However, there are no data showing the stability of 1R6F during long-term storage. The objective of this project is to study the stability of the cigarette tobacco filler and resulting mainstream cigarette smoke of 1R6F. Cigarettes were stored at -20°C, 4°C and room temperature (~22°C) for 1, 2, 3, 6, 9 and 12 months. Before they were analyzed, cigarettes stored in -20°C were transferred to 4 °C for 24hr then room temperature for at least 2hr. The reference cigarettes were conditioned for 48hr at 22°C and 60% relative humidity, prior to analysis in tobacco filler and mainstream cigarette smoke. Filler analysis included oven volatiles, alkaloids and TSNAs. Smoke analysis included the measurement of alkaloids, TSNAs, CO and TPM under ISO smoking regime. There were no significant changes for the oven volatiles, nicotine and NNN in the filler at 4°C and -20°C conditions. However, oven volatiles decreased significantly when cigarettes were stored at room temperature. There were no significant changes for puff/cigarette, CO, TPM, alkaloids and TSNAs in mainstream smoke of cigarettes under -20°C, 4°C and room temperature conditions. Our experimental data demonstrated 1R6F cigarettes were stable after one-year storage at 4°C and -20°C for selected constituents. This study will continue to monitor the stability of the 1R6F cigarette supply for the duration of this research project.

19. VALIDATION OF A LIMIT TEST METHOD FOR ANALYSIS OF PRIMARY AROMATIC AMINES (PAA) IN ELECTRONIC CIGARETTE LIQUID AND AEROSOL. Elizabeth R. ANDERSON, Jason D. Beane, Carl J. Adams and Salem Chouchane; Eurofins Lancaster Laboratories, Winston-Salem, NC, USA

Primary aromatic amines (PAAs) are compounds of interest that are routinely found in mainstream tobacco smoke. With the increasing interest in e-cigarette aerosol and liquid products, there is requirement for sensitive methods of measurement for PAAs when evaluating these matrices. The method described here is a modification of a method used for combustible cigarettes for PAA analysis in e-liquid and aerosol.

This method is a limit test, and utilizes a low level standard and a matrix spike to demonstrate realistic low level responses for comparison against samples. The low level standard is used as a clean matrix reference while the matrix spike represents a true representation of the responses of each analyte in various formulations of Electronic Cigarette e-liquid and aerosol. The target PAAs analyzed are 1-aminonaphthalene, 2-aminonaphthalene and 4-aminobiphenyl. Electronic cigarette aerosol is collected using a 44-mm Cambridge filter pad (CFP) which is then extracted with 5% HCL through both MCX and HLB cartridges. The final fraction is collected in toluene and derivatized with heptafluorobutyric anhydride. The sample is analyzed by GCMS using negative chemical ionization (NICI). The method has been validated as a limit test in concordance with the ICH guidelines.

20. SIMPLE METHOD FOR PARTICLE SIZE DISTRIBUTION DETERMINATION OF SMOKELESS TOBACCO PRODUCTS. Salem CHOUCHANE, Paul A. Lineberry and Luke D. VanHorn; Eurofins Lancaster Laboratories, Winston-Salem, NC, USA

The particle size of smokeless products such as moist snuff can be used to characterize the products. The goal of this study was to determine the particle size distribution of different smokeless products. The particle distribution was determined using a simple sieve method and was validated. Validation parameters include precision, and intermediate precision. Challenges encountered in the method validation were in determining the appropriate sieve size and the degree of moisture in the samples.

Note that the 71st TSRC had to be rescheduled due to a hurricane. As a result, several papers were withdrawn, and several had to be presented by the author's colleagu

22. VALIDATION OF A LIMIT TEST METHOD FOR BENZO(A)PYRENE IN ELECTRONIC CIGARETTE LIQUID AND AEROSOL BY LIQUID CHROMATOGRAPHY WITH FLUORESCENCE DETECTION. Darius B. GRISSOM, Carl J. Adams and Salem Chouchane; Eurofins Lancaster Laboratories, Winston-Salem, NC, USA

Benzo(a)pyrene is on the list of HPHC's in the recent FDA draft guidance for Premarket Tobacco Applications (PMTA) for Electronic Nicotine Delivery Systems (ENDS). Because this analyte is not typically detected in electronic cigarettes, a limit test offers an more efficient and cost effective analysis. A validated quantitative benzo(a)pyrene method for traditional combustible cigarettes was modified to include electronic cigarette liquid and aerosol matrices incorporating a limit test. The limit test approach reduces solution preparation, chemical cost, and analysts experimental and review time.

The method was developed using UPLC chromatographic separation through an Agilent Zorbax Eclipse XBD-C18 column, 4.6 x 50 mm, 1.8 um particle size. The method has been validated and the results of the method validation will be reported.

23. VALIDATION OF A QUANTITATIVE METHOD FOR THE TRACE LEVEL ANALYSIS OF ETHYLENE GLYCOL AND DIETHYLENE GLYCOL IN ELECTRONIC CIGARETTE LIQUID AND AEROSOL SAMPLES. Darius B. GRISSOM, Joe Kennaday, Carl J. Adams and Salem Chouchane; Eurofins Lancaster Laboratories, Winston Salem, NC, USA

USP employs a GC-FID method to test for the Limit of Diethylene Glycol and Ethylene Glycol in raw materials Prolylene Glycol and Glycerin (NMT 0.1%, each analyte). Eurofins Lancaster Laboratories has developed and validated a GCMS method that can report values down to trace levels in e-liquid (0.002% ethylene glycol, 0.003% diethylene glycol) and e-liquid aerosol (0.006% ethylene glycol, 0.003% diethylene glycol) products.

The method was developed using a GC/MS Chromatographic separation through a Restek Stabilwax column, 30m x 0.25mm x 0.25um. This poster highlights the validation parameters used to successfully validate this method, the calibration range of the method, and critical factors discovered during the validation. One specific critical factor was the successful completion of recovery studies with high product variability.

24. VALIDATED METHOD FOR THE QUANTITATIVE DETERMINATION OF CARBONYLS IN ELECTRONIC CIGARETTE AEROSOL AND LIQUID. Nancy QIAN, Jonathan Wilkins, Angela Seamans, Carl J. Adams and Salem Chouchane; Eurofins Lancaster Laboratories, Winston-Salem, NC, USA

Carbonyls are typically found in tobacco and cigarette smoke. The Food and Drug Administration have included carbonyls in the list of HPHC's in the recent draft guidance for premarket tobacco application for electronic nicotine delivery systems for both e-liquid and aerosol. The level of carbonyls in electronic cigarettes is typically found at trace levels and current methods used for analysis of carbonyls in tobacco and cigarette smoke are not sensitive enough to detect trace levels of carbonyls. As a result, Eurofins Lancaster Laboratories developed and validated a method to detect and quantitate trace levels of carbonyls in electronic cigarette e-liquid and aerosol. The carbonyls that are measured are formaldehyde, acetaldehyde, acrolein, and crotonaldehyde.

The Carbonyls are analyzed using a Waters Acquity UPLC with Waters Xevo MS/MS detector. The separation uses an Acquity UPLC BEH Shield RP18 column with 1.7 µm particle size, 100 mm length x 2.1 mm diameter. The Xevo TQ-S Triple Quadrupole with electrospray in negative ion mode and MRM is used for the detection. This method is validated to quantitate at an LOQ of 5ng/mL for Formaldehyde, Acetaldehyde, Acrolein, and Crotonaldehyde. The method validation parameters include specificity, linearity, precision, accuracy, range, robustness, solution stability and determination of LOD and LOQ.

Note that the 71st TSRC had to be rescheduled due to a hurricane. As a result, several papers were withdrawn, and several had to be presented by the author's colleague.

25. TRACE LEVEL AMMONIA ANALYSIS IN E-LIQUID AND AEROSOL PRODUCTS. Nolan D. SPANN, Jonathan Wilkins, Angela Seamans, Carl J. Adams and Salem Chouchane; Eurofins Lancaster Laboratories, Winston-Salem, NC, USA

A method was established to detect trace levels of ammonia in E-liquid and E-cigarette aerosol matrices. The new method includes both a limit test and a quantitative analysis step if ammonia levels are present above the limit test level (50 ng/mL). The method was developed using Ion chromatography. Chromatographic separation was achieved using a Dionex™ IonPac™ CS16 column, 250 x 3 mm, 5 µm particle size. The method validation results will be presented.

26. VALIDATION OF A QUANTITATIVE METHOD FOR THE ANALYSIS OF ANABASINE IN ELECTRONIC CIGARETTE LIQUID AND AEROSOL. Norman E. FRALEY, Carl J. Adams and Salem Chouchane; Eurofins Lancaster Laboratories, Winston-Salem, NC, USA

Anabasine has been identified on both the FDA HPHC list for Tobacco Products and Tobacco Smoke as well as on the list of compounds for a Premarket Tobacco Application (PMTA) for Electronic Nicotine Delivery Systems (ENDS). With the increasing interest in electronic cigarette aerosol and liquid products, there is a need for quantitating low levels of anabasine. A method was developed and validated for the determination of the presence of anabasine in electronic cigarette liquid and aerosol using liquid-liquid extraction and GC-MS-MS. Samples are dissolved in alkaline aqueous salt solution, extracted with dichloromethane and quantified based on the anabasine MRM transitions. The method validation results will be reported.

27. VALIDATION OF A LIMIT TEST METHOD FOR SELECTED VOLATILE ORGANIC COMPOUNDS IN VAPOR BY GC-MS. Norman E. FRALEY, Carl J. Adams and Salem Chouchane; Eurofins Lancaster Laboratories, Winston-Salem, NC, USA

There is interest to analyze trace levels of 1,3-butadiene, isoprene, acrylonitrile, benzene and toluene, in electronic cigarette liquid and aerosol. These analytes are not typically present in these systems, so limit tests offer a more efficient and cost effective approach for analysis. A limit test method was developed and validated for 1,3-butadiene, isoprene, acrylonitrile, benzene and toluene in electronic cigarette liquid and aerosol using gas bubbler collection followed by analysis by GC-MS. Samples are dissolved in methanol and quantified based on the compound specific SIM transitions. The method validation results will be presented.

28. A COMPARISON OF A HIGH AND LOW RESOLUTION ICP-MS FOR THE ANALYSIS OF TOXIC METALS IN PLANT MATRICES. Derek D. BUSSAN[1], Timothy L. Danielson[2] and Karl A. Wagner[2]: [1]Eurofins Lancaster Laboratories, Richmond, VA USA; [2]Altria Client Services, Richmond, VA USA

Inductively coupled plasma mass spectrometry (ICP-MS) is the analytical tool of choice for routine analysis of trace metals. For this work, we determined nine elements (As, Be, Cd, Co, Cr, Hg, Ni, Pb, and Se) in plant matrices using a high-resolution magnetic sector instrument (HR-ICP-MS) and a quadrupole instrument that was equipped with a collision and dynamic reaction cell. These elements were targeted because they are listed on the FDA's established list of Harmful and Potentially Harmful Constituents (HPHCs) for cigarettes and smokeless tobacco. As, Co, Cr, Ni, and Se are known to suffer from polyatomic and/or molecular spectral interferences which are commonly addressed using collision and/or reaction cell technology in quadrupole based systems. However, high resolution (HR)-ICP-MS instruments have the ability to resolve these polyatomic and/or molecular spectral interferences with a resolving power of 10,000 atomic mass units. The objective of this study was to compare the analytical figures of merit, i.e. limits of detection (LOD), precision, and accuracy using both types of mass spectrometers. Certified plant reference materials were used to evaluate the figures of merit. It was found that the quadrupole based instrument had superior detection limits for As and Se when using collision and reaction cell technology as compared to HR-ICP-MS due to the trade off in sensitivity when the HR-ICP-MS is operated in high resolution mode. However, the quadrupole ICP-MS had lower precision than the HR-ICP-MS for elements that do not suffer from polyatomic atomic interferences.

Note that the 71st TSRC had to be rescheduled due to a hurricane. As a result, several papers were withdrawn, and several had to be presented by the author's colleagu

41. DETERMINATION OF SELECT VOLATILE ORGANIC HYDROCARBONS IN CIGAR SMOKE. Charles BROOKS; ITG Brands, Greensboro, NC, USA

Testing for select volatile organic hydrocarbons has been in common practice for cigarette smoke for many years. Recently, testing for cigar smoke has become of potential interest. While methods for smoking and analysis for TNCO for cigars are in place, no standardized methods exist for VO determination in cigar smoke. An analytical method was developed and applied to a range of cigar styles including machine-made little cigars (large filter cigars) and cigarillos, and larger hand-rolled products. Cigars were smoked according to existing cigar smoking regimes recommended by CORESTA for TNCO analysis (CRM No. 46, CRM No. 64, CRM No. 65). During smoking, the volatile smoke components were collected in methanol contained in -70°C cold-trapped impingers, to which the pad and 200 µL of Benzene-d6 ISTD was added prior to mixing and transferring aliquot to amber GC vials for analysis. A DB-5ms column (60m x 0.25µm x 1.0µm) connected to a DB-Wax column (1.1m x 0.25µm x 0.25µm) was used for separation with a constant flow rate of 1.2 mL/min. Extracts were analyzed for 1,3-butadiene, isoprene, acrylonitrile, toluene, benzene, and styrene via gas chromatography with single quad Mass Selective Detection in EI/SIM mode. In addition to determining the levels of analytes present in smoke for a range of cigars, the impact of lighting technique and conditioning time were explored.

42. COMPARISON OF PROBABILITY OF RISK ASSOCIATED WITH CIGAR EXPOSURE. Karshak KOSARAJU, Felix Ayala-Fierro and Rob Stevens; ITG Brands, Greensboro, NC, USA

Cigars are unique tobacco products of variable physical and chemical properties. Limited information is available about exposure to smoke and probability of risk, if any, to smoke constituents. Cigar smokers can be exclusive or dual smokers of cigars and conventional cigarette products. Literature data indicates that exclusive cigar smokers do not inhale smoke into their lungs in contrast to dual smokers. Approximately one-third of total cigar nicotine is taken into the smoker's mouth as mainstream smoke and available for absorption mainly from the buccal mucosa. For dual smokers the overall exposure to a cigar is represented by both inhalation and buccal whereas for exclusive cigar smokers the overall exposure includes buccal and transdermal absorption through direct contact with lips. The objective of this study was to conduct a comparative probability risk assessment (PRA) of cigar products using cigar-specific "input" variables.

Quantitative risk assessment (QRA) of two cigar products resulted in differences of 10-30% (depending on the analyte) between products; however, probabilistic risk assessment (PRA) revealed the change was too small and may not be considered significant. In fact, PRA showed > 90% overlap between probability distributions of the two products. Sensitivity analysis suggested that smoking behavior and other exposure parameters contribute to > 90% of the observed variability associated with the analyte of concern between the products. Results overall indicates that PRA is a useful tool when comparing the probability of risk of cigar products.

43. ANALYSIS OF AROMATIC AMINES IN MAINSTREAM CIGARETTE AND CIGAR SMOKE BY GC-MS. Jeff ZHU, Charles Brooks and Lee Pittaway; ITG Brands, Greensboro, NC, USA

Polyaromatic amines have been routinely tested for cigarette smoke for several years. Typical methods employed require complex sample extraction and clean up procedures, long run times, and extensive instrument maintenance. We sought to develop an improved method for analysis of 1 & 2-aminonaphthalene and 3 & 4-aminobiphenyl with application to cigarette and cigar smoke testing. Simplified sample preparation steps employed in the 2016 CORESTA Aromatic Amines collaborative study for seven aromatic amines using a method from BAT – Souza Cruz were adopted for use with this method. Based on the referenced technique: steps for liquid-liquid extraction, neutralization, and drying and concentrating steps were eliminated. For rapid analysis of the target compounds, gc column choice and parameters were optimized to allow for a short run time while maintaining or improving selectivity and sensitivity over our previous internal method. The optimized method was demonstrated to be applicable to cigarette and cigar smoke. Results for a range of products along with validation results will be discussed. A comparison between existing methods and the new method will be provided, as well.

Note that the 71st TSRC had to be rescheduled due to a hurricane. As a result, several papers were withdrawn, and several had to be presented by the author's colleague.

44. ASSESSING POPULATION HEALTH EFFECTS OF TOBACCO PRODUCT USE FOR REGULATORY COMPLIANCE: A COMPARISON OF AVAILABLE METHODS. Florian TEISCHINGER; JT International, Geneva, Switzerland, and Bill Poland; Certara USA, Menlo Park, CA, USA

The US FDA encourages use of statistical models to project effects of a new tobacco product on population health, for Modified Risk Tobacco Product Applications (MRTPAs) and Pre-Market Tobacco Applications (PMTAs). We review different methods chosen for recently developed models. "Cohort models" follow a single cohort through the death of each member, while full ("cross-sectional") population models follow a population of mixed ages and tobacco exposures over a specified period, including births as well as deaths. Both types either calculate expected counts of members of each product use category, or use summary statistics from Monte Carlo simulation of tobacco use histories of many individuals. Output metrics include the difference in cumulative deaths between a reference scenario and a scenario where a new product is added, or may use measures such as smoking-attributable deaths or life years, perhaps quality-adjusted. Each choice has advantages; we give examples of each model type and review pros and cons.

Full cross-sectional population models have been criticized as overly complex, subject to bias, and limited to inadequately short time periods, unlike cohort models. However, we demonstrate that a full population model does not need not to have any of these limitations, and can even reduce to a cohort model if birth rates are set to zero (and initial ages are set to a single age if desired). Thus, additional complexity becomes the user's choice. In return for the birth rate requirement, full population models project annual prevalence of use of each product and mortality. This allows validation to include running over historical years to check evolution of the total population, smokers, and former smokers, all with age and sex breakdowns.

45. HIGHLY SENSITIVE NICOTINE PLASMA AND METABOLITE ASSAYS FOR CLINICAL TRIALS BY LC-MS/MS. Gene RAY, Yansheng Liu, Marsha Luna, X. Steven Yan, Julie Showalter and Kimberly Jackson; KCAS Bioanalytical and Biomarker Services, Shawnee, KS, USA

Human plasma assays for nicotine and its key metabolites (cotinine and trans-3'-hydroxycotinine) have been validated in compliance with FDA guidelines in support of clinical trials. Success in establishing these assays is careful screening of all reagents and mitigating nicotine contamination throughout the laboratory. Two separate assays were validated for nicotine. A low curve, high sensitivity assay with quantitation limit of 0.2 ng/mL and high curve assay with a quantitation limit of 0.5 ng/mL.

The low curve nicotine assay had precisions (% CV) of ≤ 7.3% CV and accuracies (% bias) of ≤ 3.9% above the LLOQ. The precision and accuracy for nicotine at the LLOQ were 12.3% and 3.9%, respectively. The high curve nicotine assay had precisions of ≤ 4.4% and accuracies of ≤ 4.9% above the LLOQ; the precision and accuracy at the LLOQ were 8.6% and 4.6%, respectively. The precision and accuracy for cotinine and trans-3'-hydroxycotinine methods were comparable to that of nicotine.

Stability was documented for each analyte in plasma for 18.5 hr at ambient temperature, following 4 freeze/thaw cycles and for up to 48 days at -20 oC. Stability was also document in whole blood for 1.5 hrs at room temperature.

Assay specificity was confirmed in the presence of menthol (as glucuronide metabolite) and in presence of lipemic and hemolytic plasma. These methods meets or exceeds FDA regulatory requirements and acceptable for clinical trial support.

47. THE IMPACT OF 2, 4-DINITROPHENYLHYDRAZINE (DNPH) BACKGROUND IN THE DETERMINATION OF CARBONYLS COMPOUNDS IN E-CIGARETTE VAPOR. Carmen DONISA, Peter Joza, Angel Rodriguez-Lafluente and William Rickert; Labstat International, Kitchener, ON, Canada

Carbonyl compounds, are known to be present in some e-liquid formulations and can be formed during aerosol generation. Analysis of these compounds, as their hydrazones (2,4-dinitrophenylhydrazine; DNPH) are routinely performed using high performance liquid chromatography (HPLC) with either ultraviolet

Note that the 71st TSRC had to be rescheduled due to a hurricane. As a result, several papers were withdrawn, and several had to be presented by the author's colleagu

(UV) or tandem mass spectrometry (MS/MS) detection. However, the variability associated with the background carbonyl levels found in the DNPH solutions, can impact the interpretation of analytical data, particularly in long term stability and simulated use studies. Differences in background carbonyl levels have been found amongst DNPH suppliers, as well as variability from lot-to-lot and container-to-container. In recent years, increased background levels of carbonyls have been found for formaldehyde, acetaldehyde, acetone and methyl ethyl ketone (MEK). Background carbonyl levels (on a μg/collection basis) have ranged from 1.4 to 12.2 for formaldehyde, and non-detected to 10.7 for acetaldehyde. When analyzing low delivery products, it is increasingly difficult to distinguish between the background levels from the DNPH solution, background associated with the vaping environment, and that generated by the product during the vaping process. Consequently, DNPH is routinely re-crystallized in the laboratory. Although this can reduce levels of some compounds like formaldehyde, other compounds such as acetone and MEK may be present in trace amounts in the solvents used during the recrystallization process, increasing the background levels. Purchasing of a custom DNPH recrystallized 5 times by the supplier, was found to have comparatively low amounts of formaldehyde and acetaldehyde. DNPH hydrochloride has been evaluated as another approach to background reduction with some success.

48. A COMPARISON OF TWO DIFFERENT EXTRACTION METHODS FOR AROMATIC AMINES IN MAINSTREAM SMOKE USING GAS CHROMATOGRAPHY-TANDEM MASS SPECTROMETRY (GC/QQQ). Andy STINSON[1], Sherry Gilliland[1], Andrew Huckins[1], Sherri S. Brown[2] and I. Gene Gillman[2]; [1]Liggett Group, Mebane, NC, USA; [2]Enthalpy Analytical, Durham, NC, USA

The established GC-MS methods for the analysis of aromatic amines (o Anisidine, o Toluidine, 2, 6 Dimethylaniline, 1 Aminonaphthalene, 2 Aminonaphthalene, and 4 Aminobiphenyl) involve complex sample preparation and analyzing the samples on a gas chromatograph with a single-quadrupole mass spectrometer. In this study, we have elected to use a triple-quadrupole mass spectrometer that greatly improves the selectivity of detection of the compounds of interest.

The objective of this study was twofold: (1) compare the establish two –step SPE extraction method verse a one-step extraction method and (2) compare the results using a triple-quadrupole mass spectrometer under electron impact ionization, a triple-quadrupole mass spectrometerr using chemical ionization and the established single-quadrupole method (GC-MS).

We will present our finding that the triple-quadrupole mass spectrometer enables simplification of the extraction method offers superior limits of detection compared to the quadrupole mass spectrometer based standard method. We will present instrument/ method precision, accuracy and LOD & LOQ of both methods.

49. DETERMINATION OF NICOTINE, NNN, AND NNK IN A NEW CIGARETTE TOBACCO FILLER STANDARD REFERENCE MATERIAL (SRM 3222) AND ITS SMOKE. Walter B. WILSON, Jeanita S. Pritchett and Lane C. Sander; National Institute of Standards and Technology, Gaithersburg, MD, USA

The United States has an annual production of approximately 800 million pounds of tobacco with estimated domestic sales of around 260 billion cigarettes/year. The National Institute of Standards and Technology (NIST) has recently collaborated with the Center for Tobacco Products at the Food and Drug administration (FDA) to develop a new low nicotine tobacco filler Standard Reference Material (SRM 3222). SRM 3222 was prepared from air-cured, low nicotine tobacco and is intended primarily for use in evaluating the accuracy of current and new analytical procedures for measuring nicotine, N-nitrosonornicotine (NNN), 4-(methylnitrosamino)-1-(3-pyridyl-1-butanone (NNK), and moisture in cigarettes sold daily. The certified values for nicotine, NNN, and NNK are based on the results of measurements made by NIST, U.S. Centers for Disease Control and Prevention (CDC), and commercial laboratories using isotope dilution liquid chromatography with tandem mass spectrometry (ID-LC-MS/MS) with different sample preparation techniques. Briefly, 0.250 g portions of finely ground SRM 3222 were vortex in a 15 mL polypropylene tubes with either 10 mL of 100 mM aqueous ammonium acetate or 5 mL of methanol extraction solution for a total of 1 h and 18 h, respectively. In the case of the smoke analysis, preliminary studies have been

Note that the 71st TSRC had to be rescheduled due to a hurricane. As a result, several papers were withdrawn, and several had to be presented by the author's colleague.

conducted using gas chromatography/MS and LC-MS/MS to identify nicotine, NNN, and NNK in the tobacco smoke collected from burning approximately 10 g of SRM 3222.

50. MENTHOL CONTENT IN CANADIAN CIGARETTES. Ivana KOSARAC, M. M. Patel, A. L. Doane, R. Scholtz, N. Mladjenovic and T. K. Mischki; Health Canada, Ottawa, Canada

In 2009, the Government of Canada prohibited the addition of flavours to cigarettes as a measure to reduce the use of cigarettes by youth. At the time this prohibition did not include menthol, a flavour known to impart cooling and analgesic properties to cigarette smoke. Recently, the sale of flavoured tobacco has been prohibited in provinces such as Alberta, Nova Scotia, and New Brunswick and the Government of Canada has published a regulatory amendment to remove menthol from its list of permitted ingredients as of October 2, 2017. Currently, menthol cigarettes represent 4.7 % of total cigarette sales in 2014. The objective of this study was to examine the menthol content and emissions from menthol cigarettes and compare this to new "green" branded cigarettes, introduced in the Alberta market after prohibition of flavourings in tobacco. Menthol content averaged 2.8 +/- 1.2 mg/g tobacco while menthol emissions averaged 0.38 +/- 0.16 and 1.1 +/- 0.30 mg/cigarette for ISO and Health Canada emissions protocols respectively. "Green" branded cigarettes had below LOD and LOQ levels of menthol in tobacco and emissions measurements. Nicotine and tar composition in tobacco and emissions varied across all brands but were comparable to previously analyzed non-menthol cigarettes.

52. BACILLUS SPP ISOLATED FROM SMOKELESS TOBACCO PRODUCTS. Casandra WEST[1] and Amy Devlin[2]; [1]RAI Services Company, Winston-Salem, NC, USA; [2]R. J. Reynolds Tobacco, Winston-Salem, NC, USA

Globally, there are an estimated 300 million users of smokeless tobacco. Since smokeless tobacco products are placed in direct and prolonged contact with the oral mucosa, researchers are beginning to evaluate the associated microbiomes of such products. Presently, there have been no reports describing the intra-species diversity of culturable (viable) bacilli isolated from smokeless tobacco and very few studies describing intra-species Bacillus diversity in general. Yet, over 90% of culturable microorganisms from tobacco products belong to the genus Bacillus. This group of organisms is ubiquitous and incredibly diverse. Bacillus species have been isolated from soil, air, sediment, aquatic and marine environments as well as from humans, cattle, fish and birds. There are over 170 species recognized to date and countless subspecies; many of which have important industrial, agricultural and medicinal uses. Several strains have been employed as probiotics for humans, livestock and aquaculture operations. Recently, marker gene amplification meta-genomic studies have been applied to evaluate these products and can support paths to more specific genetic investigations. Such studies provide valuable, all-encompassing views of the DNA of bacterial families that may be present in smokeless tobacco products; however, the presence of DNA does not necessarily indicate the presence of a living organism. Herein, this investigation focuses on the characterization of representative Bacillus strains isolated from smokeless tobacco products through evaluations of inter- and intra-species relatedness, genetic similarity and relatedness to previously studied and published strains.

53. ASSESSMENT OF FLAVOR TRANSFER TO AEROSOLS OF ELECTRONIC NICOTINE DELIVERY SYSTEMS. Courtney G. CULBERT[1], David T. Szabo[1] and Amanda G. Hudson[2]; [1]RAI Services Company, Winston-Salem, NC, USA; [2]R.J. Reynolds Tobacco, Winston-Salem, NC, USA

In the evaluation of electronic nicotine delivery systems (ENDS), reliable analytical methods are needed to acquire flavor data for consumer exposure risk assessment calculations. Without accurate transfer data, the conservative default assumption that 100% of the flavor added to the e-liquid is being transferred to the aerosol may be an overestimation. Flavor transfer values can be calculated by comparing the amount of a flavor in an e-liquid to the amount observed in a generated aerosol. These flavor amounts can be determined using both quantitative and semiquantitative gas chromatography mass spectrometry (GC-MS) methods. In the quantitative method, absolute flavor amounts are calculated using known standards, while a semiquantitative approach estimates the amount of flavor by comparing to an internal standard. Because the inherent error associated with semiquantitation of flavors in both e-liquids and aerosols is expected to be similar, the semiquantitated flavor transfer values would also be expected to be similar to the quantitated

Note that the 71st TSRC had to be rescheduled due to a hurricane. As a result, several papers were withdrawn, and several had to be presented by the author's colleagu

flavor transfer values. Testing this assumption on a variety of flavors, the flavor with the lowest transfer between e-liquid and aerosol was measured at 18% (quantitated) versus 17% (semiquantitated), while the flavor with the highest transfer was 78% (quantitated) versus 80% (semiquantitated). This trend held for all flavors investigated, suggesting that using flavor transfer data from the semiquantitative analysis would be predictive of quantitative analysis. Since semiquantitation is less labor and cost intensive than quantitation, semiquantitative flavor transfer studies are a valuable tool for the evaluation of ENDS flavors and consumer exposure.

54. CHARACTERIZATION OF INHALATION EXPOSURE TO CIGARETTE SMOKE. Charlene LIU and Kristin Marano, RAI Services Company; RAI Services Company, Winston-Salem, NC, USA

Quantitative risk assessment (QRA) may inform regulatory decisions regarding tobacco products. Evaluation of human health risks from cigarette smoking requires an adequate assessment of the exposure, which is a challenging task in part because the concentration of chemicals in the respiratory tract is not constant. No regulatory guidance currently exists for exposure assessment of tobacco products, although examples exist in the peer-reviewed literature. The U.S. Environmental Protection Agency (EPA) provides guidance that addresses methods for quantitative evaluation of exposure and risk, which is useful and can be applied to tobacco products. Two different methods were developed to quantify inhalation exposure with machine-generated smoke yields from a market survey of U.S. cigarettes. The first method treats exposure to a chemical in smoke as a continuous process and estimates an exposure concentration by averaging the yields of the chemical from cigarettes consumed over the average daily volume of air inhaled by a user. The second method treats exposure to the chemical as discrete smoking sessions and estimates a respiratory concentration of the chemical via summation of discrete smoking sessions over the course of a day. Both methods incorporate standard exposure parameters to derive an upper-bound lifetime average exposure to the chemical. For simplicity and conservatism, both methods assume 100% retention of the chemical in the smoker's body. The two methods provide risk estimates with relative percent differences within 30%; the first method was more conservative (i.e., risk-maximizing). Exposure assessment of tobacco products should be consistent with available evidence, guidance, and state of the science for risk assessment. These findings indicate that incremental modifications to exposure input assumptions do not materially affect the QRA results.

55. QUANTITATIVE RISK ASSESSMENT: THE PROPOSED NNN STANDARD. Kristin MARANO[1], P. Robinan Gentry[2], and Charlene Liu[1]; [1]RAI Services Company, Winston-Salem, NC, USA; [2]Ramboll Environ, Monroe, LA, USA

The US Food and Drug Administration (FDA) published a proposed standard for N-nitrosonornicotine (NNN) content in finished smokeless tobacco (ST) products. The proposed NNN limit of 1 part per million dry weight (ppm dw) was derived based on a target excess lifetime cancer risk (ELCR), although the rule states the increased risk of oral cancer was the impetus for the rule. A review of the quantitative risk assessment (QRA) conducted by FDA was undertaken, with findings from this review indicating deficiencies in FDA's QRA. First, the methods used to estimate the NNN cancer slope factor are inconsistent with derivation methods recommended by the US Environmental Protection Agency (EPA) and methods supported by FDA. Second, key input assumptions (i.e., body weight and lifespan) are inconsistent with current EPA recommendations in risk assessment practice. Third, the absorption factor estimated by FDA was incomplete. Fourth, the estimated ELCR in the proposed rule conveys an unrealistic level of precision, inconsistent with EPA recommendations. Finally, it is not clear that an ELCR calculation is relevant for the establishment of an NNN limit: ELCR is inadequate as a measure of excess cancer deaths in the population, because it does not account for competing mortality. However, if the ELCR is calculated in accordance with current EPA risk assessment guidance, an NNN level ≥5 ppm dw would result. This range is consistent with, or lower than, historical NNN levels in the ST products used by participants in Swedish epidemiological studies demonstrating no meaningful increase in oral cancer risk. There is no evidence that suggests setting the product standard to a ≥5-fold lower concentration of NNN will further reduce cancer risks in ST users.

56. EPIDEMIOLOGICAL EVIDENCE AND FDA'S PROPOSED NNN STANDARD. Tiffany PARMS[1], Sandra Sulksy[2], Greg Mariano[3] and Kristin Marano[1]; [1]RAI Services Company, Winston-Salem, NC, USA; [2]Ramboll Environ, Amherst, MA, USA; [3]Ramboll Environ, Arlington, VA, USA

The US Food and Drug Administration (FDA) has advanced a proposed product standard of 1μg/g N-nitrosonornicotine (NNN) content in finished smokeless tobacco (ST) products, citing selected findings from epidemiology to support the proposed limit. An independent review of FDA's application of epidemiological literature identified a number of inaccuracies. First, FDA combines oral cancer relative risk (RR) estimates for men and women, which is inappropriate given the RR between genders are widely different. Furthermore, men are the predominant users of ST. Second, FDA relied upon Swedish epidemiology to indicate current (low) levels of NNN in modern Swedish ST products are not associated with increased risk of oral cancer; however, NNN levels in Swedish ST in use during the time of the epidemiology studies were higher than levels in current products. Third, FDA relied on studies of international ST products (e.g., Asia and Africa), yet the composition and use behaviors associated with ST products unique to Asia and Africa differ markedly from those of US products, and are not relevant to US ST products and users. Fourth, NNN concentrations in products used by study participants in the available epidemiology studies cannot be estimated precisely, and there is substantial heterogeneity in the concentration of NNN and other toxicants across and within ST product types. Finally, FDA's conclusion that NNN is the predominant driver of excess oral cancer risk among ST users is inconsistent with existing scientific data, as urinary levels of NNN are generally higher among ST users compared with smokers, yet smokers incur a substantially higher risk for oral cancer than ST users. Thus, considered objectively, the available epidemiology data do not support the proposed NNN standard.

57. AKR1C1 AS A BIOMARKER FOR DIFFERENTIATING THE BIOLOGICAL EFFECTS OF COMBUSTIBLE FROM NON-COMBUSTIBLE TOBACCO PRODUCTS. Quynh T. TRAN[1], Sangsoon Woo[2], David Henderson[2], Hong Gao[3], Wolfgang Zacharias[3], Gang Liu[1] and G.L.Prasad[1]; [1]RAI Services Company, Winston-Salem, NC, USA; Axio Research, Seattle, WA, USA; University of Louisville School of Medicine, Louisville, KY, USA

Smoking has been established as a major risk factor for developing oral squamous cell carcinoma (OSCC), but less attention has been paid to the effects of smokeless tobacco products. Our objective is to identify potential biomarkers to distinguish the biological effects of combustible tobacco products from those of non-combustible using oral cell lines. Normal human gingival epithelial cells (HGEC), non-metastatic (101A) and metastatic (101B) OSCC cell lines were exposed to different tobacco product preparations (TPPs) including cigarette smoke total particulate matter (TPM), whole-smoke conditioned media (WS-CM), smokeless tobacco extract in complete artificial saliva (STE), or nicotine (NIC) alone. We performed microarray-based gene expression profiling and found 3456 probe sets from 101A, 1432 probe sets from 101B, and 2717 probe sets from HGEC to be differentially expressed. Gene Set Enrichment Analysis (GSEA) revealed xenobiotic metabolism and steroid biosynthesis were the top two pathways that were upregulated by combustible but not by non-combustible TPPs. Notably, aldo-keto reductase genes, AKR1C1 and AKR1C2, were the core genes in the top enriched pathways and were statistically upregulated more than 8-fold by combustible TPPs. Quantitative RT-PCR results statistically support AKR1C1 as a potential biomarker for differentiating the biological effects of combustible from non-combustible tobacco products.

58. THE TOXICOLOGICAL EVIDENCE FOR THE PROPOSED NNN STANDARD. Robinan GENTRY[1], Tracy Green[1] and Kristin M. Marano[2]; [1]Ramboll Environ, Monroe, LA, USA; [2]R. J. Reynolds Tobacco, Winston-Salem, NC, USA

The US Food and Drug Administration (FDA) has published a proposed standard for N-nitrosonornicotine (NNN) in smokeless tobacco (ST) products. An independent review of the toxicological evidence was conducted. Results of this review suggested that FDA failed to consider relevant scientific evidence. First, NNN concentrations associated with increased incidence of cancer in animals are much larger than expected human exposure from use of ST products. Second, differences in metabolism of NNN have been shown across species, which is important in understanding the relevance of animal results to humans. Third, there is no evidence of concordance between target tissue tumor formation across species to support

Note that the 71st TSRC had to be rescheduled due to a hurricane. As a result, several papers were withdrawn, and several had to be presented by the author's colleagu

conclusions regarding the overall toxicological evidence for oral and esophageal tumors. Fourth, esophageal tumors reported in rats, and characterized as malignant in the proposed rule, are largely benign, based on reported histopathological evaluations, and may not be representative of oral cancer in humans; when only malignant tumors are considered, the incidence of oral tumors is statistically increased in only one of five studies. Fifth, there were no repeated dose oral animal toxicity studies identified that included multiple dose treatments; therefore, there is a lack of evidence to describe the dose-response relationship between NNN oral exposure and oral cancer. Finally, in the only published study conducted according to methods comparable to Organization for Economic Cooperation and Development (OECD) guidelines for chronic toxicological assays, rats exposed to NNN in diets containing actual ST or ST extract showed no tumors of the oral cavity, esophagus, or pharynx. Based on an assessment of the available toxicological data, evidence does not support that the proposed NNN limit in ST products would be protective of public health.

59. MEASURING PARTICLE SIZE DISTRIBUTIONS OF ELECTRONIC CIGARETTE AEROSOLS UNDER VARIOUS PUFFING FLOW RATES AND DILUTION RATIOS. Chen SONG; Reynolds American, Winston-Salem, NC, USA

Particle size distributions (PSD) of electronic cigarette (e-cigarette) aerosols have lately been measured using a variety of instruments, such as electric mobility instruments, light scattering and extinction instruments and cascade impactors. These studies usually diluted the e-cigarette aerosol samples significantly to avoid exceeding the detection thresholds of the instruments. The measured e-cigarette aerosol particle size was therefore smaller due to the volatile nature of the chemical composition of e-cigarette aerosol. In addition, various puffing flow rates and e-cigarette liquid chemical compositions were used. The purpose of the current study is to systematically investigate the changes of e-cigarette aerosol PSD under different dilution ratios and puffing flow rates. The e-cigarette aerosols were generated using both ciga-like and tank type e-cigarettes. Dilution ratios ranged from 1:1 up to 1:10, while three different puffing flow rates were tested. The test matrix also included e-cigarette liquid chemical compositions prepared with different propylene glycol and glycerin ratios. A multi-angle light scattering instrument and a mini-MOUDI cascade impactor will be employed to measure the e-cigarette aerosol PSD. Results from preliminary experiments showed that e-cigarette aerosol particle size decreased with increasing puffing flow rates. Furthermore, a slight decrease in the e-cigarette aerosol particle size was observed under low dilution ratios. The findings of this study have implications for accurately measuring e-cigarette aerosol particle size and the dynamic changes of e-cigarette aerosol under various conditions such as the human respiratory tract.

60. HANDMADE PREMIUM CIGARS SMOKE EMISSIONS - LIMITATIONS RELATED TO TNCO DETERMINATION VARIABILITY. Beatrice TEILLET[1], Christian Schulz[2] and Stephane Colard[1]; [1]SEITA, Fleury-les-Aubrais, France; [2]Reemtsma Cigarettenfabriken Hamburg, Germany

To date, few studies have been published for handmade premium cigars. The assembly process of a premium cigar involves combining natural leaves which are then rolled by hand. This results in a variable product. In this study, a range of handmade premium cigars were analysed for tar, nicotine and carbon monoxide (TNCO) according to CORESTA recommended methods. These methods were initially developed for testing machine made cigars. Smoke testing of cigars by these methods was challenging and led to variable results.

Challenges were faced due to variation in the diameters of the cigars, measured at 33 mm from the mouth end after cutting. Discrepancies between calculated puff volumes of up to 3 mL per puff were observed for a given ring gauge. In addition, puff counts were highly variable, as high as double for a given product. The yield variability observed ranged from 52 to 160% for tar, 60 to 160 % for nicotine and from 32 to 120% for carbon monoxide.

As a consequence of the substantial variability when smoked, it was not possible to distinguish between different handmade premium cigars, with the exception of the TNCO value for the smallest cigar format when compared to a large cigar. The limitations of TNCO results for premium cigars and the need for method refinement and further investigations are discussed.

Note that the 71st TSRC had to be rescheduled due to a hurricane. As a result, several papers were withdrawn, and several had to be presented by the author's colleague.

61. SIMULTANEOUS DETERMINATION OF TOBACCO ALKALOIDS, TOBACCO SPECIFIC NITROSAMINES, AND SOLANESOL IN CONSUMER PRODUCTS USING UPLC-ESI-MS/MS. John R. SHIFFLETT, Leah Watson, Devin J. McNally, and Dawit Z. Bezabeh; U.S. Treasury Department, Alcohol and Tobacco Tax and Trade Bureau, Beltsville, MD, USA

The U.S. Alcohol and Tobacco Tax and Trade Bureau (TTB) is responsible for collecting Federal excise taxes on tobacco products. Tobacco products in the U.S. may fall into several taxable categories including cigars, cigarettes, snuff, chewing tobacco, pipe tobacco and roll-your-own. The existence of these taxable categories means that the TTB is also responsible for the determination of proper tax classification. Not only does proper classification determine the amount of tax owed, comprehensive classification procedures must also determine if a consumer product is subject to the tobacco excise tax. Since a product must contain tobacco to be subject to the excise tax, laboratory methods that test for the presence of tobacco can provide useful information to ascertain taxable status of a product.

To test for the presence of tobacco, an analytical method was developed that permits the simultaneous determination of nicotine and related alkaloids, tobacco specific nitrosamines (TSNA), and solanesol in methanolic extracts of tobacco. The method utilizes ultra performance liquid chromatography with electrospray ionization – tandem mass spectrometric detection (UPLC-ESI-MS/MS) and was optimized for the analysis of nicotine, cotinine, nornicotine, anatabine, myosmine, anabasine, isonicoteine, nornicotyrine, nicotyrine, N-nitrosoanatabine (NAT), N-nitrosoanabasine (NAB), 4-(methylnitrosamino)-1-(3-pyridyl)-1-butanone (NNK), N'-nitrosonornicotine (NNN) and solanesol. The analytical method was designed to attenuate the instrument response of nicotine, which is overwhelming, to permit simultaneous analysis of all analytes.

67. DEVELOPMENT OF AN ANALYTICAL METHOD TO DETECT METHYL SALICYLATE IN HUMAN PLASMA. Sandra MILLER[1], Kirk Newland[2], Jianmin Liu[1] and Mohamadi Sarkar[1]; [1]Altria Client Services, Richmond, VA, USA; [2]Celerion, Lincoln, NE, USA

Wintergreen is used as a common flavor in many moist smokeless tobacco (MST) products and is also widely used as a food flavorant as well as an over-the-counter analgesic to relieve muscle aches. MST products, including wintergreen flavored products, have been in the commercial market for decades. Unlike food, MST is not ingested and it is important to understand how much of the flavor is absorbed during use. Therefore, a method of detecting methyl salicylate (the main chemical component of wintergreen flavor) in human plasma is needed to estimate the absorption of this compound during use of wintergreen flavored MST. We report a novel high sensitivity LC-MS/MS method for the determination of methyl salicylate in human plasma. An aliquot (0.0750 mL) of human plasma containing the analyte and internal standard, d4-Methyl Salicylate is extracted using a protein precipitation with derivatization procedure. The extracted samples are analyzed by an HPLC equipped with an AB SCIEX API 4000™ triple quadrupole mass spectrometer using an ESI source. Positive ions are monitored in the multiple reaction monitoring (MRM) mode. Quantitation is determined using a weighted linear regression analysis (1/concentration2) of peak area ratios of the analyte and internal standard. The range of quantification is 1.75 -50.0 ng/mL. In a clinical study, peak plasma levels during a single 40-minute use of 2 g wintergreen flavored MST ranged from < LOQ - 4.62 ng/mL. Development of this method is an important step in being able to assess the absorption of methyl salicylate during wintergreen MST use. This method also has broader application and may be used to characterize methyl salicylate uptake from other consumer products as well.

68. THE SCIENTIFIC EVIDENCE FOR THE PROPOSED NNN STANDARD. Kristin MARANO[1], Charlene Liu[1], Geoffrey M. Curtin[1], Tiffany Parms[1], Annette Bachand[2], Sandra I. Sulsky[2], Greg Mariano[2], Tracy Greene[3] and P. Robinan Gentry[3]; [1]RAI Services Company, Winston-Salem, NC; USA, [2]Ramboll Environ, Amherst, MA, USA; [3]Ramboll Environ, Monroe, LA, USA

The US Food and Drug Administration (FDA) published a proposed standard of 1 μg/g (dry weight) N-nitrosonornicotine (NNN) in finished smokeless tobacco (ST) products. FDA's rationale for the standard is that NNN is a potent carcinogen, NNN is a major contributor to the elevated cancer risks associated with ST use, and establishing such a limit is appropriate for the protection of public health. However, a public health benefit associated with the proposed NNN limit is not supported by the available scientific evidence, which includes toxicological and epidemiological data, results of quantitative risk assessment (QRA), and modeling projections of population mortality based on potential changes in use behaviors. Specifically, animal experimental designs, including tested exposure concentrations and observed outcomes, are not relevant to humans. Historical levels of NNN in Swedish products used to estimate health risks, and which demonstrate no meaningful risk of oral cancer, were well above the proposed standard. FDA's QRA approaches do not conform with US Environmental Protection Agency recommendations in risk assessment practice, and results are implied to be representative of oral cancer risk in humans when they are not. Finally, changes in use behaviors likely to be associated with the proposed standard - and the resulting impact on overall population health - are not appropriately considered by FDA. Realistic scenarios for use behaviors likely to result from the proposed standard (e.g., increased use of cigarettes) suggest an increase in mortality that offsets the benefits proposed. Based on an independent evaluation of the epidemiological and toxicological evidence, QRA, and likely effect on population health, the proposed limit of NNN in ST products would not benefit, or otherwise be protective of, public health.

69. CRITICAL ASSESSMENT OF THE ANALYTICAL METHODS MEASURING NNN IN SMOKELESS TOBACCO PRODUCTS. Candice K. CUNNINGHAM; RAI Services Company, Winston-Salem, NC, USA

In January 2017, the U.S. Food and Drug Administration (FDA) issued the proposed rule, "Tobacco Product Standard for N-nitrosonornicotine (NNN) in Finished Smokeless Tobacco Products." Within the proposed rule, FDA recommended an analytical method for measuring NNN in smokeless tobacco products (with provisions for "alternative methods"). The FDA's laboratory information bulletin (LIB), "Determination of N-nitrosonornicotine in Smokeless Tobacco and Tobacco Filler by HPLC-MS/MS" referenced CORESTA Recommended Method (CRM) 72, an established, consensus method for analyzing smokeless tobacco products to quantify NNN.

The release of the proposed standard, including the LIB, highlights the need for readily available, robust, high-throughput analytical methods for determination of NNN in order to comply with regulatory requirements. This presentation will discuss the figures of merit and the challenges of analytical methods for determining NNN in the smokeless tobacco matrix as it relates to the proposed product standard and analyses currently performed in the tobacco industry. Discussion will include sample handling protocols, analytical instrumentation selection, moisture versus water content determination, and dry weight basis NNN calculations. FDA's LIB will be compared to well-established methods such as CRM 72, CRM 76, and those available through the Centers for Disease Control. Additionally, best practices regarding robust method development will be assessed (ASTM, FDA draft guidance, etc.) as these relate to our shared understanding of future proposed standards in a regulated industry.

70. GENETIC APPROACHES AND RESOURCES FOR ENABLING COMPLIANCE TO PROPOSED NNN STANDARDS. Ralph E. DEWEY[1], Ramsey S. Lewis[1], Jianli Lu[1], Lucien Bovet[2], Simon Goepfert[2] and Prisca Campanoni[2]; [1]North Carolina State University, Raleigh, NC, USA; [2]Philip Morris International, Neuchatel, Switzerland

Tobacco-specific nitrosamines (TSNAs) are formed through the nitrosation of tobacco alkaloids. Over the past decade, we have demonstrated how altering the genetics of the tobacco plant can result in dramatic reductions in TSNA formation. This has been accomplished by reducing either the levels of the alkaloid precursor of a given TSNA, or the nitrosating agent involved. By introducing, then combining, mutations in the three nicotine demethylase genes responsible for production of the NNN precursor nornicotine, we have observed reductions in the levels of nornicotine and NNN in air-cured tobacco leaves by as much as 80% in comparison to varieties lacking these mutations. As an alternative strategy, we have demonstrated that the expression of a constitutively active nitrate reductase enzyme can mediate substantial reductions in all classes of TSNAs within the leaves of air cured plants. In addition to reviewing these previously published means of minimizing NNN accumulation, we will also present the results of a new strategy of TSNA reduction. By inhibiting the expression of a specific nitrate transporter gene, we observed an approximate 66% reduction in NNN within air-cured leaves. An additional advantage of the latter two strategies for TSNA reduction is that along with NNN, substantial reductions are observed in the levels of NNK, another TSNA that could be the subject of mandated reductions by the FDA in the future. The

Note that the 71st TSRC had to be rescheduled due to a hurricane. As a result, several papers were withdrawn, and several had to be presented by the author's colleague.

utilization of tobacco varieties containing gene mutations that greatly decrease the propensity for NNN production likely represents the most effective means for ensuring compliance to NNN standards while retaining important leaf quality characteristics.

71. NNN LEVELS IN STABLE REDUCED CONVERTER (SRC) AND LC LINES CURED UNDER CONDITIONS THAT FAVOR NNN FORMATION. Marcos LUSSO, Kenworth Lion, Andrew Adams, Whit Morris, Ujwala Warek and James Strickland; Altria Client Services, Richmond, VA, USA

N-nitrosonornicotine (NNN), a tobacco specific nitrosamine (TSNA) found in cured leaf, filler and smoke, is formed by nornicotine nitrosation. We previously demonstrated a ~75% reduction of cured leaf NNN content in burley and dark tobaccos having stable reduced convertor status (SRC) compared to Low Converter plants (LC). Here we present NNN data for burley and dark tobacco SRC and LC lines cured under normal conditions (control) and under conditions favourable for TSNA formation (experimental). SRC and LC burley were housed in a control barn and air cured according to recommended practices. Additionally, an experimental burley barn was packed with about ~1/3 more SRC and LC burley tobacco plants than the control barn and managed to retain high relative humidity during curing. For dark tobacco, SRC and LC lines were housed in a control barn and fire cured according to recommended practices. An experimental Dark barn was packed with SRC and LC Dark tobacco and cured at temperatures up to 173°F having open flames during curing. At the end of curing, leaf samples were collected and analyzed for TSNAs. SRC plants had ~63% and 64% less NNN than LC plants in control barns for burley and dark tobacco, respectively. NNN levels increased in SRC and LC tobaccos when cured in experimental barns compared to control barns. However, SRC NNN levels in experimental barns were ~76 and 61% lower relative to LC plants cured under identical conditions for burley and dark tobacco, respectively.

72. N-NITROSONORNICOTINE REDUCTION IN DARK TOBACCO VARIETIES AND SMOKELESS PRODUCT PROTOTYPES. Marcos LUSSO, Andrew Adams, Benjamin Lewis, Timothy Poyner, Kenworth Lion, Lionel DeLoach, Timothy Danielson, James Franke, Ujwala Warek and James Strickland; Altria Client Services, Richmond, VA, USA

During the past decades there has been a substantial effort to reduce the level of the tobacco specific nitrosamine N-nitrosonornicotine (NNN) and its precursor nornicotine in tobacco and tobacco products. Research on the mechanism of nornicotine formation led to the identification of three tobacco genes (CYP82E4, CYP82E5 and CYP82E10) encoding for cytochrome P450 nicotine demethylases that convert nicotine to nornicotine. Through conventional breeding, we developed dark tobacco varieties (Stable Reduced Converter/SRC varieties) containing the three non-functional nicotine demethylase genes. Tobacco varieties containing this new technology, named ZYVERTTM technology, were grown on on-farm experiments and, leaf produced from these varieties, were used to make smokeless product prototypes. Control product prototypes were produced using leaf from commercial low converter (LC) varieties. Our results indicate that the presence of non-functional alleles of the three genes reduces the level of nornicotine and NNN in dark tobacco varieties by about 70% compared to the level observed in commercial LC dark tobacco varieties. The reduced levels of NNN in the cured leaf of SRC varieties was also observed in smokeless product prototypes.

73. COMPARISON OF INTRA-CROP YEAR VARIABILITY IN NNN IN TOBACCO AND NNN LEVELS IN SMOKELESS TOBACCO PRODUCTS. Michael J. OLDHAM, David J. Phillips, Jason L. Jordan, Michael J. Morton, Kenworth E. Lion III, Marcos F. Lusso, James E. Franke and James A. Strickland; Altria Client Services, Richmond, VA, USA

Although tobacco is an agricultural crop, little has been published on the variability in NNN levels in tobacco grown within a single crop year (intra-crop year variability) by multiple growers. Furthermore, the potential impact of intra-crop year variability of NNN levels in tobacco on NNN levels in smokeless tobacco products is unknown. In this work, NNN levels were measured in dark air-cured, dark fire-cured and burley tobaccos used to produce smokeless tobacco products over either a six or ten crop year period. Additionally, NNN levels were measured in nine smokeless tobacco products manufactured over seven years. Only results from smokeless tobacco products manufactured using a single crop year of tobacco were included. NNN measurements over ten crop years for dark air-cured and dark fire-cured tobacco and

Note that the 71st TSRC had to be rescheduled due to a hurricane. As a result, several papers were withdrawn, and several had to be presented by the author's colleagu

over six years for burley tobacco demonstrated a 10 fold range (lowest concentration to highest concentration) within a single crop year. Mean NNN levels exceeded 1ppm in dark-fire cured and burley tobacco in every year tested and 6 of 10 years for dark-air cured tobacco. Depending on the specific smokeless tobacco product and year, the measured NNN levels generally varied from three to six fold within a single crop year. Over the seven years, mean NNN levels (dry weight basis) exceed 1ppm for the vast majority of smokeless tobacco product measurements. The demonstrated variability in measured levels of NNN in smokeless tobacco products, due to natural variability in levels of NNN in tobacco, provides significant insight about measured NNN levels among smokeless tobacco products.

74. THE ENANTIOMERIC COMPOSITION OF N'-NITROSONORNICOTINE IN AIR-CURED TOBACCO INFLUENCED BY NICOTINE DEMETHYLATION. CAI Bin[1], Huihua Ji[2], Franklin F. Fannin[2], and Lowell P. Bush[2]; [1]Guizhou Academy of Tobacco Science, Guiyang City, China; [2]University of Kentucky, Lexington, KY, USA

N′-Nitrosonornicotine (NNN) is a potent and organ-specific carcinogen found in tobacco and tobacco smoke in substantial amounts. The aim of this study was to investigate the variations in NNN enantiomeric composition in nicotine demethylase mutants. A gas chromatography/thermal energy analyzer method using two columns in series was developed to separate the enantiomers of NNN, N′-nitrosoanabasine, and N′-nitrosoanatabine. Tobacco lines with different combinations of three nicotine demethylases inhibited were grown in the field. Air-cured leaves were analyzed for the concentration and enantiomeric composition of nicotine, nornicotine and NNN. In mutant lines, the concentration of NNN ranged from 0.44 µg g-1 to 13.63 µg g-1 and the percentage of (R)-NNN of total NNN varied from 7% to 69%. The measured NNN had the same enantiomeric composition as nornicotine, rather than nicotine, even when the level of nornicotine was reduced to 0.6% of nicotine in a triple mutant line. The pattern of the enantiomeric composition of nicotine, nornicotine, and NNN demonstrated that the direct formation of NNN from nicotine, if it occurs, is negligible in air-cured tobacco. Since (S)-NNN is dominate form in smokeless tobacco and more carcinogenic than its R form, the reduction of (S)-nornicotine should be a priority for the reduction of NNN.

76. THE POWER OF TACTILITY: TIPPING PAPER IN INTERACTION WITH THE CONSUMER. Michael LINDNER and Eike Schopper; Tannpapier, Traun, Austria

Tipping Paper plays a highly essential role for the manufacture of filter cigarettes as it connects the filter plug with the tobacco rod, controls the level of smoke yields via Tipping perforation and acts as design tool for the customized branding of cigarettes. However, besides these technical and visual aspects, Tipping is also the only component of a cigarette which is in direct contact with the lips of consumers. Since the human lips are rather sensitive to mechanical and physical surface properties of touching objects, Tipping Paper offers perfect opportunities to interact with smokers by stimulating the tactile senses of the labial areas. Hereby, the first idea is to use the lip-release effect of commercially applied Tipping Papers to generate a comfortable impression during the smoking process. While lip-release by means of transparent varnishes or color coatings is a state-of-the-art feature, the recently introduced Super Lip-Release Tipping comprises superior hydrophobic characteristics for an outstanding smoking experience for different smoking habits. In this context, the water-repellent strength of Super Lip-Release and standard Tipping Paper will be demonstrated and compared with a mathematical absorption model and a small survey amongst regular smokers. The second approach to communicate with the consumer is to activate the haptic perception on the lips and fingers. Based on a sophisticated embossing technology, Textured Tipping represents a smart way of realizing an extraordinary cigarette mouthpiece which comprises unique haptic surface patterns on its Tipping Paper. Using various Textured Tipping samples, potential application and design options as well as structural limits will be thoroughly discussed. As a conclusion, Super Lip-Release and Textured Tipping are powerful examples to indicate the fusion of technical functionality and appealing tactility.

Note that the 71st TSRC had to be rescheduled due to a hurricane. As a result, several papers were withdrawn, and several had to be presented by the author's colleague.

77. AN APPROACH TO IDENTIFY FACTORS RELATING TO CHANGES IN CHEMICAL CONSTITUENTS IN CURED TOBACCO LEAF DURING AGING. Atsushi NAGAI and Daichi Mochizuki; Japan Tobacco, Yokohama, Japan

The aging process, the long-term preservation process which takes place before cigarette production, improves the aroma/taste compared with non-aged tobacco leaves. In attempts to understand the reaction mechanism of chemical changes in the aging process, many studies have been conducted. Three factors (chemical reaction, enzymatic reaction and microbial action) have been thought to be causative in the changes. However, it is still unclear which factor is dominant in the chemical changes, because of the difficulty in distinguishing between the effects of the three factors. This research intended to identify dominant factors relating to chemical changes during aging. We report on: (i) preparation of model samples that could distinguish between the three factors, (ii) an aging test using the model samples, and (iii) identification of dominant factors related to changes in each chemical constituent during aging.

Flue-cured Virginia, cultivated and cured in Japan, was used for this test. Samples of the leaves were pre-treated with super-heated steaming (SHS) and/or electron beam irradiation (EB) before the aging test. SHS preferentially deactivated enzyme rather than microbe. In contrast, EB successively sterilized, without enzyme deactivation. All of the samples were aged simultaneously at 22°C / 60% RH for 6 months, or at 33°C / 60% RH for 1 month. The pre-aging/post-aging changes in several chemicals including sugars, organic acids, amino acids, and alkaloids, were observed. The dominant factors relating to the changes in each constituent could be estimated from the differences in the pattern of change between model samples. Results indicated that most of the changes in the constituents analyzed in this experiment were produced mainly by chemical reaction, or by the combination of chemical reaction and enzymatic reaction.

78. DEVELOPMENT OF A MULTI-STAGE LIQUID EXTRACTION METHOD FOR THE QUANTITATIVE ANALYSIS OF FLAVORS IN TOBACCO FILLER. John A MATHIS, R. A. Stanelle and J.-M. D. Dimandja; U.S. Food and Drug Administration, Atlanta, GA, USA

A method to quantify flavor chemicals in the filler of combustible tobacco products was validated using GC-MS. The tobacco filler samples from cigarettes, little cigars, cigarillos, and cigars were prepared in a multi-step extraction procedure using ethyl acetate and water followed by dilution with hexane. The features of the extraction stages include: extraction of non-polar compounds into ethyl acetate, water addition for polar compounds, partitioning into two-layers with subsequent centrifugation, followed by the removal of the ethyl acetate layer and mixing with an equal volume of hexane containing the internal standard (IS), second centrifugation, yielding an extract for GC-MS analysis. The selected flavor chemicals are quantitated by GC-MS in full scan mode by relating the peak area ratios to the concentration using target compound extracted ion chromatograms (EIC) compared to the IS peak area.

The linearity, accuracy, precision, limit of quantitation (LOQ), and selectivity results were used to demonstrate the extraction and instrumental analysis was appropriate for the quantitative analysis. The linearity of the method was assessed across the range of 10 to 300 ppm (μg/g). The accuracy and precision was evaluated using replicate measurements of matrix spiked samples at 50, 100, and 150 ppm levels, in addition to the LOQ standard at 10 ppm. The LOQ was established by evaluating the precision of all compounds, $<$ 10% relative standard deviation (%RSD), as well as the signal-to-noise ratio for the lowest absolute peak area EIC, $S/N \geq 11$. The accuracy ranged from 95-127% for individual replicates and the precision at each concentration level ranged from 3-8% RSD. The selectivity results, which required no interfering peaks in the blank matrix and sufficient resolution at the LOQ level, met the validation criteria by visual interpretation and illustrated for each EIC. In addition to the quantitative analysis of the selected flavor compounds, the use of the full scan mode can also provide identification and library searching capabilities for unknown compounds. The results from the validation support the use of the proposed method for quantitative analysis of flavor related compounds in unburned portion of combustible tobacco product filler.

Note that the 71st TSRC had to be rescheduled due to a hurricane. As a result, several papers were withdrawn, and several had to be presented by the author's colleagu

79. PHYSICAL CHARACTERISTICS OF HANDMADE PREMIUM CIGARS: SPECIFICATIONS AND CONSEQUENCES OF SPECIFICATIONS. Beatrice TEILLET, Thomas Verron, Stephane Colard; SEITA, Fleury-les-Aubrais, France

Handmade premium cigars consist of natural and heterogeneous leaves rolled by hand to a given dimension. This traditional making process is performed by people (rollers) with considerable experience and training. The rollers adapt their technique, cigar to cigar, according to the leaves' natural characteristics, such as volume or shape during bunching. This traditional manual process needs to be taken into account when assessing the physical characteristics of a cigar.

This study investigated diameter, length, weight and pressure drop of handmade premium cigars. A wide range of sizes were sampled at one point in time and analysed. In order to assess their respective variability, a distinction was made between those characteristics related to cigar specification and those related to the consequence of a specification.

Results showed low variability of diameter and length, below 10% and 2% respectively, whereas weight and pressure drop exhibited substantially higher variability up to 40% and 120% respectively. This confirms that diameter and length, i.e. specified parameters, are directly and well controlled conditions. In contrast, weight and pressure drop are not directly controlled but are simply a consequence of the cigar specifications and the manual making process.

80. WASTE REDUCTION IN THE PRIMARY BY USING MICROWAVE IN-LINE TECHNOLOGY FOR PROCESS CONTROL AND OUTGOING GOODS INSPECTION. Hendrik RICHTER; TEWS Elektronik, Hamburg, Germany

Meeting quality control objectives in the Primary encompasses the processing of tobaccos at client-target moisture levels within associated control ranges for each blend. To achieve these quality goals, it is crucial that process control be effectively managed and that final quality inspection can be verifiably executed. Currently, but certainly in the past, such verifications and control functions were usually performed by intermittent sample taking from both the conveyer line and the final inspection station. Typically, C48 boxes were sampled to provide validation or reference values from laboratory-based instrumentation. These procedures are time consuming and do not allow for automated process control at locations such as the DCC or the final dryer.

In this presentation we demonstrate that tobacco waste can be measurably decreased when deploying TEWS inline systems based on microwave resonance technology by significantly reducing in-process moisture variability. TEWS inline systems measure real-time and high-speed true core moisture while simultaneously making available associated outputs that allow for moisture control automation. TEWS microwave transmission technology provides high accuracies during the moisture inspection of bulk goods at the final inspection station.

81. A PRODUCT SURVEY OF 25 PIPE TOBACCOS PURCHASED FROM WEB-BASED MERCHANTS. John H. LAUTERBACH; Lauterbach & Associates, Macon, GA, USA

In May 2016, the FDA extended their regulatory authority to cover pipe tobaccos and other tobacco products, which were not originally included in the FSPTCA. Pipe smokers are a small percentage of US tobacco users, but they have a wide variety of products to smoke ranging from expensive traditional products to much lower cost products that can be found in many retail locations that sell cigarettes. However, there is little knowledge about blends, casings, flavors and the processing that are unique to pipe tobaccos, particularly the more expensive products that are sometimes in non-cut forms and often have special processing that is atypical for cigarette tobaccos. Consequently, we conducted a product survey of 25 pipe tobaccos that we purchased from web-based merchants. These products included those that were twists and pressed cakes and included two products having deer tongue in the blend. In addition to analytical results from commercial laboratories, we determined cocoa, coumarin, vanillin, ethyl vanillin, licorice, polyphenols, and sugars using a gradient HPLC system consisting of 3 Waters 510 pumps, Waters 680 gradient controller, Rheodyne 7125 injector, Waters 486 TAD, and HP 3396 Series II integrator.

Note that the 71st TSRC had to be rescheduled due to a hurricane. As a result, several papers were withdrawn, and several had to be presented by the author's colleague.

Columns used were a Clippeus C18, 10 μm 4.6 x 250 mm and a YMC Triart C18, 120Å, 5 μm 250 x 4.6 mm. Methods used were adapted from those in the literature and/or publicly available. Much useful data for distinguishing among the samples came from application of scan techniques using a MeOH extract of the tobacco and a slow gradient from 93/5/2 H2O/MeOH/HOAc to 10/88/2 H2O/MeOH/HOAc and setting the TAD wavelength at 340 nm, 280 nm, or 254 nm.

83. THERMAL DEGRADATION STUDIES OF ELECTRONIC CIGARETTE LIQUIDS PART 1: A NOVEL ANALYTICAL METHOD TO STUDY A-DICARBONYL FORMATION. Matt S. MELVIN, Karen C. Avery, Regina M. Ballentine, William P. Gardner and Karl A. Wagner; Altria Client Services, Richmond, VA, USA

The formation of carbonyl compounds in electronic cigarette (e-cigarette) aerosols from thermal decomposition processes has been well established in the literature. These thermal decomposition products are thought to originate from the primary e-liquid components: propylene glycol (PG) and glycerin (GLY). The presence of specific α-dicarbonyl compounds has also been reported in e-cigarette aerosols. The α-dicarbonyl compounds of interest include glyoxal, methylglyoxal, 2,3-butanedione (diacetyl) , and 2,3-pentanedione (acetyl propionyl). The formation of these compounds is not readily explainable through typical dehydration and auto-oxidation pathways. The objective of this work is to develop an understanding of the potential reaction pathways for the formation of these compounds. To facilitate this study, an analytical method specific for α-dicarbonyl compounds was developed utilizing o-phenylenediamine (OPD) as the derivatizing agent to produce the corresponding quinoxaline product. OPD has the advantage of forming a single, stable product instead of a mixture of isomers that are susceptible to the reversible reactions that are seen when derivatizing with 2,4-dinitrophenyl hydrazine. The derivatization reaction occurs rapidly in water and is amenable for the collection of aerosols. Sample preparation and aerosol collection procedures were optimized for e-liquids and e-cigarette aerosols. Prepared samples were analyzed by liquid chromatography-mass spectrometry. This method was validated and deemed fit for purpose for the determination of the analytes of interest in e-liquids and aerosols. The optimized method conditions and the results of the validation will be presented.

84. THERMAL DEGRADATION STUDIES OF ELECTRONIC CIGARETTE LIQUIDS PART 2: DEVELOPMENT OF A MODEL REACTION SYSTEM USED TO STUDY A-DICARBONYL FORMATION. Matt S. MELVIN, Karen C. Avery, Regina M. Ballentine, William P. Gardner and Karl A. Wagner; Altria Client Services, Richmond, VA, USA

The formation of carbonyl compounds in electronic cigarette (e-cigarette) aerosols from thermal decomposition processes has been well established in the literature. These thermal decomposition products are thought to originate from the primary e-liquid components: propylene glycol (PG) and glycerin (GLY). The presence of specific α-dicarbonyl compounds has also been reported in e-cigarette aerosols. The α-dicarbonyl compounds of interest include glyoxal, methylglyoxal, 2,3-butanedione (diacetyl) , and 2,3-pentanedione (acetyl propionyl). The formation of these compounds is not readily explainable through typical dehydration and auto-oxidation pathways. The objective of this work is to develop an understanding of the potential reaction pathways for the formation of these compounds. A derivatization method using o-phenylenediamine was used to study the formation of the α-dicarbonyl compounds in e-liquids. To this end, a model reaction system that simulates a potential reaction environment of an e-cigarette atomizer was developed using a microwave reaction system. The effect of the e-liquid components (PG, GLY, nicotine, water, and flavors), reaction temperature, and time on α-dicarbonyl formation were determined. The implication of these results on potential reaction pathways for the formation of α-dicarbonyl compounds in the e-cigarette aerosolization process will be discussed.

85. THE DETERMINATION OF DIACETYL AND ACETYLPROPIONYL IN AEROSOLS FROM ELECTRONIC SMOKING DEVICES USING GAS CHROMATOGRAPHY TRIPLE QUAD MASS SPECTROMETRY. Serban MOLDOVEANU, Amanda G. Hudson and Andrew Harrison; R.J. Reynolds Tobacco, Winston-Salem, NC, USA

A reliable and sensitive method for the measurement of the levels of diacetyl (2,3-butanedione) and acetylpropionyl (2,3-pentanedione) in the aerosol (both the particles and the suspending gas) of electronic

Note that the 71st TSRC had to be rescheduled due to a hurricane. As a result, several papers were withdrawn, and several had to be presented by the author's colleagu

smoking devices (e-cigarettes) has been developed. The method uses a gas chromatographic separation on a Carbowax type column with the measurement of the analytes on a triplequad mass spectrometer working in positive MRM mode. The method has been validated using standard requirements regarding selectivity, sensitivity, recovery, accuracy, and repeatability. The limit of quantitation (LOQ) for the method was determined to be 0.41 ng/mL for diacetyl and 0.21 ng/mL for acetylpropionyl as measured for standards. These values translate to an LOQ of 0.082 ng/puff for diacetyl and 0.042 ng/puff for acetylpropionyl as measured for 50 puffs from an e-cigarette placed in 10 mL acetone. The samples analyzed included collected aerosols from several e-cigarettes, and a number of liquids used in electronic cigarettes (e-liquids). 3R4F Kentucky reference cigarette was also analyzed for evaluating the accuracy of the procedure, with good agreement with data from the literature. Diacetyl and acetylpropionyl were distributed in both particulate phase and also in vapor phase. The levels of diacetyl and acetylpropionyl in particulate phase collected from 3R4F cigarettes were found to represent only about 22% for diacetyl and only 31% for acetylpropionyl, while the vapor phase for diacetyl represented 78% and for acetylpropionyl 69% of the total analyte. The levels of diacetyl and acetylpropionyl in the aerosols of most electronic smoking devices was found to be very low. The analysis of the two analytes in e-liquids showed a very large range of levels.

86. ANALYSIS OF (S)- AND (R)-NICOTINE IN VARIOUS NICOTINE SAMPLES AND IN E-LIQUIDS. Serban MOLDOVEANU[1], Gary M. Dull[2], Ross J. Oden[1], Wayne A. Scott[1]; [1]R.J. Reynolds Tobacco, Winston-Salem, NC USA; [2]RAI Innovation, Winston-Salem, NC, USA

Present study measured the level of nicotine in 38 commercially available liquids used in commercial electronic smoking devices (e-liquids), estimated if the nicotine in these e-liquids is predominantly (S)-nicotine (obtained from tobacco), and measured the ratio of (S)-nicotine and (R)-nicotine in four different nicotine samples of USP purity (United States Pharmacopeia). The quantitative analysis of (total) nicotine in e-liquids was performed using a GC-FID technique. In addition to nicotine, the levels of propylene glycol (PG), glycerin, and menthol were determined. The analysis of the ratio of the (R)- and (S)-enantiomers was performed by two different techniques. One technique was based on a GC/MS with the separation on two columns in series RT-GammaDEXsa 30 m x 0.25 mm i.d., with 0.25 um film. The other technique was a HPLC-UV with the separation on a Chiracel OJ-3 column 250 mm x 4.6 mm with 3 um size particle. Three of the analyzed nicotine samples were obtained from the tobacco plant (Nicotiana tabacum, Nicotiana rustica, etc.) and one was synthetic nicotine. The natural nicotine from all three samples was found to be made from about 0.67% (R)-nicotine and 99.33% (S)-nicotine. The synthetic nicotine sample contained 50/50 (S)- and (R)-nicotine. All evaluated e-liquid samples contained predominantly (S)-nicotine indicating they contain natural nicotine.

87. ANALYSIS OF ORGANIC ACID CONTENT IN E-LIQUIDS AND E-CIGARETTE AEROSOL BY HIGH PERFORMANCE LIQUID CHROMATOGRAPHY – ULTRAVIOLET DETECTION (HPLC-UV). Angel RODRIGUEZ-LAFUENTE, Romulo Lodevico, Mehran Sharifi, Peter Joza and William Rickert; Labstat International, Kitchener, ON, Canada

Weak organic acids can be used as preservative agents, potentially affecting the stability of e-cigarette solutions. However, they can also impact characteristics like the aroma and taste of aerosols. The objective of this study was to develop and validate a simple analytical method for the analysis of 6 organic acids (formic, acetic, glycolic, lactic, benzoic and oxalic) possibly found in e-liquids.

Initially, one method was developed for the determination of benzoic acid. The aerosol was collected onto a filter pad and extracted with isopropyl alcohol, then diluted with a 0.1% phosphoric acid solution. The extract was then analyzed by HPLC-UV (230nm) using a Poroshell 120 EC-C18 column (2.7 μm, 100 x 4.6 mm). E-liquids were analyzed using the same chromatographic conditions as the aerosols. Limits of quantitation (LOQ's) were 17.2 μg/collection for aerosol and 1.7 μg/g for e-liquids, with recoveries of 112% and 102% for laboratory fortified matrices, respectively. To extend the method to include additional organic acids, a Supelcogel C-610H column (9 μm, 300 x 7.8 mm) was used. Although satisfactory LOQ's were obtained, ranging from 0.43 μg/g to 11.8 μg/g for e-liquids and 1.71 μg/collection to 47.3 μg/collection for aerosols, the method was hindered by the limited capacity of the column handle a higher organic content mobile phase required to elute the glycerol and propylene glycol sample matrix from the column. To overcome these long run times, a third chromatographic approach was developed using an

Note that the 71st TSRC had to be rescheduled due to a hurricane. As a result, several papers were withdrawn, and several had to be presented by the author's colleague.

Allure Organic Acids column (5 μm, 300 x 4.6 mm), able to utilize a mobile phase with up to 50% of acetonitrile (ACN), reducing the run time to 40 minutes.

88. NICOTINE PHARMACOKINETICS OF ELECTRONIC CIGARETTES: EXPERIMENTAL DATA AND A REVIEW OF THE LITERATURE. Ian M. FEARON[1], Alison Eldridge[1], Nathan Gale[1], Mike McEwan[1], Paul Nelson[2], Elaine Round[2] and Mitch Stiles[2]; [1]British American Tobacco (Investments) Limited, Southampton, UK; [2]RAI Services Company, Winston-Salem, NC USA

E-cigarettes are battery-powered electronic devices from which users can inhale nicotine following its aerosolisation from a heated liquid solution. Some regulators and public health bodies consider e-cigarettes as potentially playing a major role in tobacco harm reduction. The ability of e-cigarettes to deliver nicotine to smokers in a manner and form generally similar to cigarette smoking have been recognised as key factors in helping smokers reduce or cease the use of combustible cigarettes. Nicotine pharmacokinetic studies of e-cigarettes have been performed for several years and are beginning to show how nicotine delivery is evolving as the products themselves evolve. In this presentation, we provide a critical overview of the literature to describe what is known about nicotine delivery from e-cigarettes, by both presenting data from our own clinical studies and from what has been published in the literature. We will discuss how the progression of e-cigarette design, development, and user familiarity and subsequent use behaviour has allowed increases in nicotine delivery, in the context of how much and how rapidly nicotine is delivered during acute-use periods. This presentation will also provide insight into current research gaps, and highlight a potential need for standardisation of the methodologies used to assess nicotine uptake to facilitate comparisons between different products and between different sub-categories of e-cigarettes.

89. AN ALTERNATIVE APPROACH TO QUANTITATIVE METHODS: LIMIT TEST FOR ANALYSIS OF DEEMING ANALYTES IN ELECTRONIC CIGARETTES. Salem CHOUCHANE, Darius B. Grissom and Carl J. Adams; Eurofins Lancaster Laboratories, Winston-Salem, NC, USA

Electronic cigarettes generate an aerosol produced from a heating element powered with a battery and a liquid in a cartridge without combustion. The liquid typically consists of humectants, water, flavors and nicotine. Except for the main components of the electronic cigarette liquid and aerosol, the levels of HPHC analytes in the aerosol or liquid of electronic cigarettes are extremely low and in many cases not detectable. Performing routine and high throughput quantitative analysis for analytes that are consistently reported below the level of quantitation is costly and inefficient. Given the need to see very low levels of analytes in products where the analyte is not expected to be present, limit tests offers an opportunity to be more efficient and cost effective. Limit tests are also an acceptable testing convention of the FDA, ICH, and USP.

Eurofins Lancaster Labs developed and validated limit test methods. Our efforts have been focused primarily on modifying previously validated methods or developing new methods to include electronic cigarette liquid and aerosol matrices. This presentation will highlight the advantages of limit test methods versus quantitative methods. The limit test approach reduces solution preparation, chemical cost, and analysts experimental and review time. Several examples of validated limit test methods will be presented.

90. MARKET ENDS E-LIQUIDS AND OPEN DEVICE CHEMICAL EVALUATION. David T. SZABO[1], Amanda G. Hudson[2], and Courtney G. Culbert[1]; [1]RAI Services Company, Winston Salem, NC USA; [2]R. J. Reynolds Tobacco, Winston Salem, NC, USA

Refillable or open electronic nicotine delivery systems (ENDS) have grown in popularity within recent years. Evaluating the interaction between e-liquid ingredients and device structural materials during intended use of the product is a routine stewardship practice to mitigate any potential unexpected exposures to the consumer. Market evaluations indicate that fruit flavored e-liquids is a top selling category for e-liquids used in open ENDS. In this study, we measured several compounds in popular brands of fruit flavored e-liquids and generated aerosols from a popular open ENDS. Aerosolization used a common machine puffing regime with a 55 mL puff volume, 30 second interpuff and 3.3 second puff duration (0.3 seconds added to ensure push button actuation). E-liquid and aerosol measurements included carriers that are commonly used, as well as flavors, carbonyls, and metals. Results indicated that formation of carbonyls

Note that the 71st TSRC had to be rescheduled due to a hurricane. As a result, several papers were withdrawn, and several had to be presented by the author's colleagu

and presence of metals were dependent on e-liquid formulation interaction with the device. These findings support the need for investigation of the combinatorial response between flavoring compounds and ENDS device functional parameters, ultimately leading to the protection of public health.

91. INVESTIGATION OF TSNA FORMATION IN ELECTRONIC CIGARETTE LIQUIDS AND AEROSOL. Xiaohong Cathy JIN, Karen C. Avery, Georgios D. Karles, William P. Gardner, Matt S. Melvin and Karl A. Wagner; Altria Client Services, Richmond, VA USA

Electronic cigarette (e-cigarette) formulations typically contain tobacco derived nicotine and therefore, may contain other tobacco related components such as trace levels of tobacco specific nitrosamines (TSNAs), nitrite, and minor alkaloids. Previous reports including work in our laboratories, suggest that TSNAs are present in e-cigarette liquids (e-liquids) and aerosols. The objective of this work was to investigate the possible formation of TSNAs in e-liquids and aerosols. There are reports in the literature that nitrite can react with nicotine and minor alkaloids in tobacco and cigarette smoke to form TSNAs. However, there are no literature reports on the effect of tobacco related compounds on the formation of TSNAs in e-liquids and aerosols. These studies were conducted by fortifying (spiking) nicotine containing e-liquids (mixture of propylene glycol/glycerin/water/nicotine) with nitrite, nitrate, ammonia, and minor alkaloids. The fortified e-liquids were analyzed to assess if TSNAs were formed. These e-liquids were used to fill e-cigarettes to determine the transfer and potential formation of TSNAs during the aerosolization process. Model systems were used to provide an understanding of the TSNAs formation pathways. The experiments demonstrated that nitrite led to the formation of TSNAs in e-liquids and e-cigarette aerosol. This information should be useful for regulators and manufacturers when making science-based decisions on which HPHCs and constituents to monitor in e-vapor products.

92. COMPARISON OF RESULTS FROM THREE TECHNIQUES FOR THE ESTIMATION OF E-LIQUID pH-VALUES. John H. LAUTERBACH; Lauterbach & Associates, Macon, GA, USA

Several techniques have been reported for the estimation of the pH-values of undiluted e-liquids. Such pH-values may be useful for QA and regulatory purposes. However, they may not be relevant to the aerosol formed when the e-liquid is vaped and the nicotine is absorbed by the vaper's mouth and airways. Consequently, we compared the pH-values of two noncommercial e-liquids, using three techniques. The first involved the estimation of the pH-value after dilution of the e-liquid with water (Lisko et al., 2015). The second involved use of apparatus similar to that used for puff-by-puff smoke pH to measure the pH-values of the aerosol generated from the e-liquid. The third involved use of a glassmouth containing 10mL Pickering Laboratories 1700-0304 artificial saliva (pH ~6.8). The glassmouth was also fitted with a pH probe to measure aerosol pH. Aerosols were generated with standard V2 battery sections and blank cartomizers containing about 500mg of e-liquid. The first e-liquid was 50mg/mL nicotine in PG. The second was made by adding propionic acid (30mg, 0.4mmole) to 6mL of the 50mg/mL nicotine in PG (300mg, 1.85mmole total). The pH-values obtained after dilution of the e-liquids with water (two sample replicates per sample) were 9.56, 9.66, 6.34, and 6.26, respectively. The measured pH-values of the aerosol using the puff-by-puff apparatus were [pH-values at the 25th puff (55/3/30 puffing regimen)] were 7.99, 7.80, 6.61, and 6.48. The aerosol pH-values at the 50th puff (55/3/30) were 7.63 and 6.66 using the glassmouth. The saliva pH-values were 7.49 and 7.40; and there was little difference in the saliva nicotine concentrations. Additional e-liquids have been studied, and the results will be reported in the presentation.

93. IMPACT OF CIGAR PHYSICAL VARIABILITY ON CIGAR EXPOSURE USING PROBABILISTIC RISK ASSESSMENT. Felix Ayala-Fierro, Karshak KOSARAJU and Rob Stevens; ITG Brands, Greensboro, NC, USA

Cigars are unique tobacco products of a wide variety of sizes: length, diameter and weight, and include large cigars, small cigars (cigarette-like) and cigarillos. The weight of cigars can range widely which influences exposure to cigar constituents, either by inhalation or through direct contact with lips. Another variable is that little cigar and cigarillo smokers are typically dual users of cigars and conventional cigarette products. In addition, unlike cigarette smokers, cigarillo smokers commonly practice partial smoking and re-lighting. Cigarillo consumption also varies widely from as few as one per week to daily which

Note that the 71st TSRC had to be rescheduled due to a hurricane. As a result, several papers were withdrawn, and several had to be presented by the author's colleague.

determines the total smoking time (days/year). The objective of this study was to determine the impact of cigar physical variability on uncertainties associated with exposure.

Consumer exposure to cigars, displayed as lifetime average daily intake (LADI) was explored using probabilistic risk assessment (PRA). First, "input" variables for different exposure scenarios were established. These parameters were found to differ from traditional cigarette exposure. Second, the impact of cigar weight, cigar consumption (weekly) and smoking behavior (whole vs. partial), on the overall exposure was investigated by PRA sensitivity analysis. Results from PRA analysis indicates that the size of the cigar (weight) has the greatest impact (~50%) on the overall exposure to the analyte of concern followed by the exposure duration and cigars smoked per week. It appears that smoking behavior (whole vs. partial) has little impact on lifetime exposure from cigar exposure. For dual smokers the combined exposure from cigars and cigarettes determines the overall exposure to a given analyte of concern. Understanding the impact of cigar physical variability on the overall consumer exposure to cigars is critical to explain uncertainty associated with smoking.

94. CONCEPT OF CHARACTERIZATION AND CERTIFICATION OF THE 1R6F REFERENCE CIGARETTE. Socrates Jose P. CAÑETE; University of Kentucky, Lexington, KY, USA

The quality and reliability of data generated by validated methods are strongly reinforced by the use of reference materials (RMs) and, in particular, certified reference materials (CRMs). The 1R6F is designed to mimic a matrix that closely resembles currently marketed cigarette products in the United States. The planned production inventory is such that it will allow consistent distribution to stakeholders over five years or longer. This paper discusses the concept of characterization of the 1R6F discussing the sampling as well as the statistical treatment of analytical data from contracted laboratories. The statistical processing of analytical data involved a two-tier approach to better reflect the two-tier concept of characterization of the 1R6F. The two-tier approach reflects two uncertainties; one is the expected uncertainty that can be expected for single-laboratory use of the RM and the other is the expected uncertainty when two or more laboratories compare their data on the 1R6F. The certification of these two uncertainties allows for individual labs to set their upper and lower limits in their quality control monitoring while allowing for assessments of results (on the 1R6F) between labs.

95. ALTERNATE MATERIALS AND THEIR POTENTIAL IMPACT ON HPHCS. Michael J. MORTON, Nripa T. Jain, Kathleen H. Fox, Raquel M. Olegario, and Timothy L. Danielson; Altria Client Services, Richmond, VA, USA

It is commonplace to use alternate materials from different suppliers interchangeably in the manufacture of cigarettes. To be used interchangeably, these alternate materials will often meet the same specification, perform the same in the finished product, and result in the same finished product specifications even though the materials may not be chemically identical. The objective of this study was to evaluate the impact, if any, on HPHC yields from the use of several different types of alternate non-tobacco materials in PM USA cigarette products.

The alternate materials used in this study are all commercially available materials and included two plug wraps, two base tipping papers, two cigarette seam adhesives, two tipping adhesives, and three filter tows. To evaluate the various combinations of alternate materials, the study employed an adaptation of a fractional factorial using 16 combinations of the alternate materials. Using a designed experiment allowed us to estimate the effects of the alternate materials with much greater precision than a "one factor at a time design." Each of the cigarettes in the study used the same base design with matched tobacco filler and cigarette paper, so that the products were the same except for the alternate materials under study.

The cigarettes were tested for the smoke constituents listed in the FDA abbreviated HPHC list both under ISO and Health Canada Intense smoking regimes in ISO 17025 accredited laboratories. The results of this study demonstrate that the use of these alternate materials does not impact the HPHC yields of the cigarette product. The estimated differences in HPHC yields were all numerically small, and, after adjusting for testing multiplicity, none of the estimated differences were statistically significant.

Note that the 71st TSRC had to be rescheduled due to a hurricane. As a result, several papers were withdrawn, and several had to be presented by the author's colleagu

96. AN EVALUATION OF THE VARIABILITY OF HPHCS IN CIGARS AS COMPARED TO CIGARETTES. Karl A. WAGNER, Tammy L. Blake, Matt S. Melvin, Michael J. Morton and Jennifer H. Smith; Altria Client Services, Richmond, VA, USA

In May 2016, the U.S. Food and Drug Administration (FDA) issued a final rule to deem cigars to be subject to the Federal Food, Drug, and Cosmetic Act (the FD&C Act), as amended by the Family Smoking Prevention and Tobacco Control Act (Tobacco Control Act). As part of this regulation, the FDA will require manufacturers to report the quantities of Harmful and Potentially Harmful Constituents (HPHCs) in cigar filler and smoke. The diversity of the cigar product category adds challenges to the measurement of HPHCs. CORESTA has developed recommended methods for nicotine and CO for cigars; however, standardized methods do not exist for other HPHCs. Consensus methods, with defined repeatability and reproducibility (r&R), do exist for many of the abbreviated list HPHCs in cigarettes. However, these methods have not been shown to be fit for purpose for the analysis of cigars and we found that some of the methods developed for cigarettes need modification to be suitable for cigars. We will present data that demonstrate that HPHC variability in cigar testing is greater than that observed for cigarette testing. This presentation will cover some of the factors that affect cigar testing variability including smoke collection, variation in physical properties, and tip styles. In order to effectively test cigars for HPHCs and define the uncertainty of the reported results, the three pillars of effective analytical testing need to be implemented: consensus standardized methods with defined r&R, proficiency testing, and reference products.

97. EVALUATION OF MAINSTREAM CIGARETTE SMOKE COMPONENTS YIELDS OF DIFFERENT CIGARETTE TYPES BASED ON PRINCIPAL COMPONENT ANALYSIS. LI Zhonghao, Yang Fei, Liu Shanshan, Fan Ziyan, Wang Ying, Bian Zhaoyang, Deng Huimin, Tang Gangling and Hu Quingyuan; China National Tobacco Quality Supervision & Test Center, Zhengzhou, China

The feasibility of using Principal Components Analysis (PCA) in order to analyze a complex cigarette smoke mixture, seven cigarette mainstream smoke components, including carbon monoxide (CO), 4-(methylnitrosamino)-1-(3-pyridyl)-1-butanone (NNK), Benzo(a)pyrene (B[a]P), crotonaldehyde, phenol, hydrogen cyanide (HCN) and ammonia, were determined from 21 representative commercial cigarette samples. The test results were verified by Kaiser-Meyer-Olkin (KMO) measure of Sampling Adequacy and Bartlett' s Test of Sphericity initially, and PCA was used to identify yield patterns, clustering of different types of cigarettes and these clusters causes. The results demonstrated that the selected three Principal Components (PC) could explain about 81.5% of total data variability. Ammonia, NNK and phenol were the main contributors for PC1, crotonaldehyde, HCN and CO for PC2, and B[a]P for PC3. Based on the statistical results of the selected three PCs' scores, the description and explanation on different types of cigarette specimens (tobacco blend, design tar yield, slim, charcoal filter) were achieved successfully. Clustering of lower mainstream smoke components emission were found for slim types, and yields of seven components in mainstream smoke for the non-menthol and menthol subpopulations showed few significant differences. The study demonstrates that application of PCA to describe and analyze the dataset of complex cigarette smoke mixtures can be feasible and practical.

98. OXIDATIVE STRESS AND OXIDATIVE DAMAGE IN HUMAN LUNG EPITHELIAL CELLS INDUCED BY TOTAL PARTICULATE MATTER EXTRACTIONS FROM THREE DIFFERENT TOBACCOS. KANG Yu, Qiao Liangiun, Hu Qiao, Hua Chenfeng, Xie Fuwei, Shang Pingping, Li Xiang and Zhao Junwei; Zhengzhou Tobacco Research Institute, Zhengzhou, China

Global sales and popularity of e-cigarettes have been growing rapidly in the past few years, but there is still a big share of conventional cigarettes and other tobacco products, which had led to the concerns about the toxicities of different tobacco products. Four representational tobacco products, including two conventional cigarettes, one e-cigarette product and one snus pouch, were selected to investigate the oxidative stress and oxidative damage induced by total particulate matters (TPM) extractions. Mainstream smoke/aerosols of 3R4F, CM8 cigarettes and one e-cigarette were captured according to ISO regime with smoking machines and the extraction of CRP1 Swedish style snus pouch (mentioned as TPM) was prepared by extracting the soluble substances with DMSO. Cell viability, intracellular reactive oxygen species (ROS) and DNA oxidative damage (8-OHdG) were tested on Beas-ab cells.

Note that the 71st TSRC had to be rescheduled due to a hurricane. As a result, several papers were withdrawn, and several had to be presented by the author's colleague.

Results showed that: Extractions of cigarette smoke led to lower cell viability as compared with snus pouch, and that of the e-cigarette almost showed no effect on the cell viability within the tested dose range; Within the 0-50 μg/mL TPM doses range of the present experiment, extractions of the cigarette smoke induced significant oxidative stress to Beas- 2b cells, which also showed dose-response relationship, whereas that of the snus pouch showed significant oxidative stress when the doses were as high as 100-500 μg/mL. The extractions of e-cigarette did not show significant effects on oxidative stress induction; Under the given experimental conditions, extractions of cigarette smoke, e-cigarette and the snus pouch induced the highest 8-OHdG levels at 20, 40, 500 μg/mL exposure doses, respectively. The characteristics of oxidative damage to DNA of the four products are different from these characteristics of those to cell viability and oxidative stress to Beas-2b cells. The results had shown different *in vitro* toxicological characteristics of these tobacco products.

99. SIMULTANEOUS DETERMINATION OF 22 AMADORI COMPOUNDS IN TOBACCO LEAVES BY LC-MS/MS. WANG Xiaoyu, Liu Leiyu, Wang Sheng, Qin Yaqiong, Jia Yunzhen, Cui Huapeng, Pan Lining, Xie Fuwei and Liu Huimin; Zhengzhou Tobacco Research Institute, Zhengzhou, China

Amadori compounds are important aroma precursors. Therefore, rapid and corrective determination of Amadori compounds is meaningful for monitoring the quality of tobacco leaves. By the end of 2016, the publicated methods have analyzed no more than 6 Amadori compounds simultaneously in tobacco. And in 2017, our research group reported a HPLC- MS/MS method to simultaneously quantify 10 glucose derived Amadori compounds in tobacco. This paper is based on our previous study. With respect to another 12 Amadori compounds without commericial standards, model reactions were conducted. And by using HPLC-Q-TOF-MS with high mass resolution, the 12 Amadori compounds were identified in the model reaction systems based on being matched with accurate mass and confirmed with the fragmentation way. The quantitative and the qualitative ion pairs in MRM mode were also selected. Then using HPLC-MS/MS, we quantifyed the 10 Amadori compounds with standards using external standard quantitative method, and semi-quantified the 12 Amadori compounds without standards using the standard curves of others with similar retention time and abundance in tobacco. Through large sample test, except 4 compounds which were unstable or of abundance close to LOQ, the intra-day precisions of 18 Amadori compounds were lower than 10%. The inter-day precisions of 14 Amadori compounds were lower than 10%, and that of 4 compounds could be controlled between 10% and 15%. The developed method was used in testing tobacco leaves of different types. The results showed that the total weights of 22 Amadori compounds accounted for 2.4% in flue-cured tobacco, 1.7% in oriental tobacco, 0.14% in burley tobacco. And the content distributions of the 22 Amadori compounds were also different among different type of tobacco leaves.

100. ELECTRONIC CIGARETTE AEROSOL DYNAMICS IN A PHYSICAL MODEL OF THE ADULT HUMAN ORAL/PHARYNGEAL CAVITY. Weiling LI[1], Qiang Wang[2]; Nicolas Castro[1], Jingjie Zhang[1], Yezdi Pithawalla[1], Michael Oldham[1], Peter Lipowicz[1] and Ali Rostami[1]; [1]Altria Client Services Richmond, VA, USA; [2]Lancaster Laboratories, c/o Altria Client Services, Richmond, VA, USA

The objective of this work is to generate experimental data to validate a computational fluid dynamic (CFD) model for e-cigarette aerosol deposition. An adult human oral/pharyngeal wet walled hollow physical model has been developed for this purpose. The physical model was generated using a 3D printer from the CT scan of a 28 year-old healthy male and had an internal volume of 69.8cc. The wall was covered with a layer of cotton cloth that can be saturated with water to replicate the high humidity conditions typically encountered in a human oral/pharyngeal cavity. The model was placed in an oven at 37 °C, and measurements were taken under both wet and dry wall conditions. Deposition efficiency from a MarkTen® product using a prototype formulation was determined by measuring cumulative aerosol mass from five puffs (gravimetric) and individual constituents from a single puff (GC/MS analysis) at the entrance and exit of the physical model. Humidity at the exit of the physical model was maintained at > 90% at a constant air flow rate of 0.66 L/min. A 37 °C dry wall condition with a constant flow of 0.66 L/min through the model resulted in a mean aerosol mass loss of 6.6 ± 0.9% due to the deposition to the wall. Under wet wall conditions, the aerosol mass increased by 38% for a 5s puff duration, with 55cc puff

Note that the 71st TSRC had to be rescheduled due to a hurricane. As a result, several papers were withdrawn, and several had to be presented by the author's colleagu

volume and 37 °C wall temperature. The increase is due to moisture uptake by the aerosol. The aerosol mass increased by 80% using a 3s puff duration, 55cc puff volume and 37 °C wet walled condition. The experimental data will be used to validate the CFD model.

103. THE MEASUREMENT OF ADDICTIVENESS AND ATTRACTIVENESS. Neil SHERWOOD; Neil Sherwood Consulting, Commugny, Switzerland

Despite a rapid increase in regulatory demands for information regarding the addictiveness and attractiveness of tobacco or consumer nicotine products, there is a notable lack of clarity on how these concepts should be assessed and which metrics which should be employed. This presentation considers the core scientific constructs which underlie addictiveness and attractiveness and suggests an approach which may go some way to satisfying regulatory demands. An explanatory model is described, built around definitions and demonstrations of addictiveness and attractiveness which have been offered by various scientific bodies. The model provides a framework upon which to understand the relationship between attractiveness and addictiveness and how these may influence nicotine and tobacco use. Accordingly, addictiveness can be assessed by measures of "liking" whereas attractiveness can be assessed by measures of "wanting". Finally existing metrics are placed within this model, their properties discussed and the opportunities for development of new metrics are considered.

104. ADAPTING TRADITIONAL HUMAN ABUSE LIABILITY TESTING TO TOBACCO PRODUCTS. Graham K. Wood, and Michael M.C. KONG; Altasciences Clinial Research, Laval, QC, Canada

Under the Family Smoking Prevention and Tobacco Control Act that grants the FDA with full regulatory jurisdiction over all tobacco products, companies pursuing market access of novel tobacco products, including electronic cigarettes and smokeless tobacco products, require a premarket tobacco application (PMTA). As detailed in the FDA's draft guidance concerning PMTAs, one of the key issues that must be addressed is the risk for addiction and abuse potential. This includes the likelihood that a novel tobacco product leads non-smokers to adopt the new tobacco product and/or its effectiveness in curbing existing smokers from using combustible cigarettes. These additional regulatory demands on tobacco products draw parallels to the human abuse potential (HAP) issues that pharmaceutical companies face when developing drugs that target the central nervous system. The methods used to demonstrate HAP - for example, acute dose effect comparisons, physical dependence measures, and behavioural economic procedures, have evolved over the past few decades and are designed to estimate the liking and wanting of the product. While some experience has been gained applying some of these methods in preclinical tobacco research, there is significantly less research using laboratory based human models of tobacco use. Furthermore, the model of addiction for tobacco products is different than pharmaceuticals, due to a much higher contribution of wanting over liking to the addictiveness, requiring consideration of this in the HAP study designs. The purpose of this lecture will be to describe how some of the methods used to study HAP of drugs can be applied to tobacco products and to discuss some of the characteristics and challenges unique to tobacco.

105. A CLINICAL SITE PERSPECTIVE: THE USE OF E-SCALE VERSUS PAPER SCALE TO CAPTURE SUBJECTIVE MEASURES FROM TOBACCO/NICOTINE WITHDRAWAL AND PRODUCT EFFECT QUESTIONNAIRES. (Presented by Doug Copeland) Jonathan Austin, Lorraine Rusch, Mark Pearson and Margarita Nunez; High Point Clinical Trials Center, High Point, NC, USA

Capturing data in an accurate and timely fashion is important in clinical research trials. Visual analog scale (VAS) assessment is often used to evaluate participants' subjective responses to Potential Reduced Exposure Products (PREPs). Advances in technology are enabling researchers to move from traditional, paper-based data collection to an e-tablet based methodology.

Methods: Two clinical trials each involving three different tobacco products with 24 subjects used either paper forms or an e-tablet to investigate the abuse liability potential of given products with two questionnaires: VAS Tobacco/Nicotine Withdrawal Scale and the VAS Product Effects Questions. Both studies had similar protocol design, inclusion/exclusion criteria, and patient demographics. Data was

Note that the 71st TSRC had to be rescheduled due to a hurricane. As a result, several papers were withdrawn, and several had to be presented by the author's colleague.

collected on the following: 1) time to complete the surveys and perform quality assurance 2) number of errors detected during quality assurance.

Results. With paper assessments, there was an average of 8.9 errors per subject involving measurement of the VAS line or documenting time/date. The e-tablet based surveys had four queries in total. The mean times to complete the paper VAS/ per subject were 40.1 seconds for the Tobacco/Nicotine Withdrawal Scale and 38.8 seconds for the Product Effects Questionnaire. The e-tablet questionnaires were completed in 38.2 seconds and 36.9 seconds respectively. Mean query time by staff on the paper surveys was 3 minutes and 18 seconds. The e-tablet did not require any query time from the site perspective.

106. A STUDY TO EXAMINE CHANGES IN EXPOSURE TO CIGARETTE SMOKE CHEMICALS WHEN A SMOKER SWITCHES TO USING A TOBACCO HEATING PRODUCT. PART I: STUDY DESIGN. Nathan GALE[1], Mike McEwan[1], Alison Eldridge[1], Oscar M. Camacho[1], John McAughey[1], James Murphy[1], Chuan Liu[1], Christopher J. Proctor[1], Ian M. Fearon[1], Neil Sherwood[2], Edward Bowen[3], Simon McDermott[3] and Emma Holmes[3]; [1]British American Tobacco (Investments) Ltd, Southampton, UK; [2]Neil Sherwood Consulting, Commugny, Switzerland; [3]Covance, Leeds, UK

A tobacco heating product (THP), glo™, which heats rather than combusts tobacco and produces significantly reduced machine yields of chemical toxicants compared to combustible cigarettes, was assessed in this study. Biomarker of exposure (BoE) studies are an important tool in determining whether this reduction in toxicant emissions translates to actual reductions in human exposure to cigarette smoke toxicants when smokers switch to using glo™. Data from such studies may potentially be required as part of a regulatory package, particularly as one aspect of a modified-risk assessment of a novel product.

This presentation (Part I of II) will outline the design of a two-centre, in-clinic (confinement), forced-switching, randomised controlled clinical study, performed in Fukuoka, Japan (UMIN000024988, ISRCTN14301360) evaluating this novel THP. In this study, we compared baseline levels of selected BoE to cigarette smoke toxicants in the exhaled breath and urine of 180 smokers, to those seen when the smokers either remained smoking combustible cigarettes, switched to using a THP, or quit all tobacco use completely, for 5 days. We also assessed levels of two biomarkers of biological effect before and after switching, and determined nicotine pharmacokinetics for the study products. The study was approved by a local Institutional Review Board and was run in accordance with ICH-GCP. Subjects provided written informed consent prior to study participation and were deemed healthy following medical examination and clinical laboratory screening. Smoking status was verified by exhaled CO and urinary cotinine measurements.

107. A STUDY TO EXAMINE CHANGES IN EXPOSURE TO CIGARETTE SMOKE CHEMICALS WHEN A SMOKER SWITCHES TO USING A TOBACCO HEATING PRODUCT. PART II: STUDY FINDINGS. Nathan GALE[1], Alison Eldridge[1], Mike McEwan[1], Graham Errington[1], James Murphy[1], Christopher J. Proctor[1], Ian M. Fearon[1], Simon McDermott[2], James Glew[2] and Andrew Hedge[2]; [1]British American Tobacco (Investments) Ltd, Southampton, UK; [2]Covance, Leeds, U.K.

A tobacco heating product (THP), glo™, which heats rather than combusts tobacco and produces significantly reduced machine yields of chemical toxicants compared to combustible cigarettes, was assessed in this study. Biomarker of exposure (BoE) studies are an important tool in determining whether this reduction in toxicant emissions translates to actual reductions in human exposure to cigarette smoke toxicants when smokers switch to using glo™. Data from such studies may potentially be required as part of a regulatory package, particularly as one aspect of a modified-risk assessment of a novel product. This presentation is the second of two and will describe the findings from this study, the design of which was outlined in the first presentation.

180 subjects completed the study. Subjects who switched from smoking to exclusive use of the glo™ device (n=30) for 5 days showed reductions in levels of exhaled carbon monoxide and a range of urinary BoEs, compared to their levels at baseline. In contrast, the continued smoking group (n=30) showed little change in BoE levels throughout the study. We also present data for the continued menthol smoking group

Note that the 71st TSRC had to be rescheduled due to a hurricane. As a result, several papers were withdrawn, and several had to be presented by the author's colleagu

(n=30), for subjects switched to the glo™ menthol variant (n=30), and for those subjects who ceased all product use for 5 days (n=30).

Our data show that, when switching from smoking conventional cigarettes to using the glo™ THP, smokers experienced substantial reductions in exposure to smoke toxicants. Longer-term BoE and BoBE studies may demonstrate, respectively, that these reductions continue or are sustained beyond the duration of this study, and whether these reductions in exposure translate to reductions in smoking-related health risks in subjects who switch to using glo™.

108. COMPREHENSION AND PERCEPTIONS OF REDUCED RISK INFORMATION FOR CAMEL SNUS. Kimberly FROST-PINEDA[1], Janine Pillitteri[2], Saul Shiffman[3], and Geoffrey Curtin[1]; [1]RAI Services Company, Winston-Salem, NC, USA; [2]Pinney Associates, Burlington, VT, USA; [3]Pinney Associates, Pittsburgh, PA, USA

A continuum of risk exists among tobacco products, whereby non-combusted products present less health risks than combusted products. Achieving tobacco harm reduction requires risk information be effectively communicated to consumers, while also communicating appropriate cautions. This study assessed consumers' comprehension and perceptions of reduced risk information for Camel Snus. An online sample of 4,924 U.S. adult tobacco users and non-users viewed an advertisement, stating that smokers who switch completely from cigarettes to Camel Snus can greatly reduce their risk of lung cancer, oral cancer, respiratory disease and heart disease. Respondents answered questions about the risks of Camel Snus relative to cigarettes, absolute risks of Camel Snus, and other information in the advertisement. Across the four diseases mentioned, <10% of respondents indicated Camel Snus presented no risk at all; most (52-62% across diseases) indicated Camel Snus presented less risk than cigarettes, but still some risk. Some indicated Camel Snus had the same risk as cigarettes; this varied by disease, even though risk information did not indicate differences. Thirty-seven percent of respondents indicated the risk of oral cancer was the same for Camel Snus and cigarettes, compared to only 20% for lung cancer, suggesting that responses were shaped by intuitions and pre-existing beliefs, as well as the advertisement's content. High proportions of respondents (>80%) indicated Camel Snus is addictive, quitting tobacco use is the best choice, and non-tobacco users should not use Camel Snus – all statements included in the advertisement. Only 4% believed smokers would receive a health benefit if they continued to smoke while using Camel Snus. These findings suggest that appropriate information regarding the relative risks of tobacco products can be effectively communicated to consumers.

109. REDUCED RISK INFORMATION FOR CAMEL SNUS: PROJECTING LIKELIHOODS OF USE AMONG CURRENT SMOKERS, FORMER SMOKERS AND NEVER TOBACCO USERS. Geoffrey CURTIN[1], Karen Gerlach[2] and Saul Shiffman[2], [1]RAI Services Company, Winston-Salem, NC, USA; [2]Pinney Associates, Pittsburgh, PA, USA.

A risk differential exists among tobacco products, whereby non-combusted products present less health risks than combusted products. Informing smokers of the lower risks associated with using a reduced risk product could encourage switching to that product, and thus benefit population health. This benefit must be weighed against the possibility that such information may increase tobacco use among non-users and/or reduce quitting among smokers. Using an online sample of 11,302 U.S. adults, likelihood of use for Camel Snus was assessed among current and former smokers, and never tobacco users. Respondents were randomized to view an advertisement stating that smokers who switch completely to Camel Snus can greatly reduce their risk of lung cancer, oral cancer, respiratory disease and heart disease (while also conveying cautionary information); or, a control advertisement that only described Camel Snus. Respondents indicated their intent to purchase Camel Snus for trial (1-10 scale); these ratings were then converted to projected likelihoods of use, via a predictive algorithm. Projected Camel Snus use was significantly higher among current smokers than former smokers or never tobacco users. Importantly, the reduced risk information differentially increased projected use among current smokers (8.2%), while having a minimal effect on former smokers (1.9%) and never tobacco users (0.5%). For never tobacco users, projected use was highest among those already susceptible to tobacco use. For current smokers, projected use was lower among those likely to quit (4.2%) than those not likely to quit (8.7%). These findings suggest that providing reduced risk information for Camel Snus may increase use among current smokers not likely to quit, with minimal potential to increase use among groups who could be harmed by adopting the product.

110. MODELING THE POPULATION HEALTH EFFECTS OF CAMEL SNUS WITH REDUCED RISK INFORMATION. Annette Bachand[1], Sandra Sulsky[1], Saul Shiffman[2] and Geoffrey CURTIN[3]; [1]Ramboll Environ U.S., Amherst, MA, USA; [2]Pinney Associates, Pittsburgh, PA, USA; [3]RAI Services Company, Winston-Salem, NC, USA

We present findings from a statistical model, used to assess the population health effects of Camel Snus with reduced risk information. Analyses followed a single cohort of 1 million males from age 13 to age 72, comparing the number of survivors in a base case (cigarette use only) to a counterfactual scenario (cigarette or Camel Snus use). Age-specific likelihoods that relevant groups – smokers who otherwise would quit, smokers who otherwise would continue smoking, non-tobacco users who otherwise would initiate smoking, and non-tobacco users who otherwise would remain non-users – would use Camel Snus with reduced risk information were derived from consumer testing that projected purchase probabilities; secondary transitions, such as potential gateway effects for Camel Snus, used conservative probabilities. Analyses that considered all beneficial and harmful transitions estimated that increased Camel Snus use would improve survival by ~7,000 individuals (assuming 89-92% risk reduction, versus smoking); and, a survival benefit would be retained if Camel Snus reduced risk by as little as 55%, compared to smoking. Extrapolation of the estimated survival benefit to a realistic U.S. cohort of 4.1 million (both genders), suggests survival to age 72 years would be increased by >25,000 individuals. Tipping point analyses that included all primary and secondary harmful transitions indicated that if ~1.5% of continuing smokers switched completely and persistently to Camel Snus at each age interval, there would be a survival benefit. These single cohort analyses demonstrate that the transition with the greatest effect on population health is complete switching among smokers who otherwise would continue smoking. Collectively, these analyses indicate that providing reduced risk information for Camel Snus is likely to have a beneficial effect on overall population health.

111. FLAVORED E-CIGARETTE USE AMONG U.S. ADULTS: RESULTS FROM TWO NATIONAL SURVEYS. Mimi KIM[1], Mark Sembower[2], Saul Shiffman[2]; and Geoffrey Curtin[1]; [1]RAI Services Company, Winston-Salem, NC, USA; [2]Pinney Associates, Pittsburgh, PA, USA

Flavored e-cigarette use among U.S. adults was examined, based on data from two national surveys. PATH provides a representative sample of 32,320 adults (1,575 current established e-cigarette users), surveyed in-person from September 2013 to December 2014; the National Tobacco Behavior Monitor (NTBM) provides a weighted sample of 46,637 adults (4,845 past-30-day e-cigarette users), surveyed online from January 2014 to June 2015. The surveys define current e-cigarette use differently, yet provided similar findings. Overall, two-thirds of e-cigarette users reported using flavored varieties (PATH: 67%/NTBM: 67%). Flavored e-cigarette use was lowest among non-Hispanic Caucasians (63%/62%). In NTBM, African Americans reported the highest rate of flavor use (86%), due to a high percentage using menthol e-cigarettes; 78% of African Americans in PATH reported using flavored varieties, but flavor was not specified. Flavor use declined steadily with age in both surveys, from 83%/85% among 18-24-year-olds to 50%/41% among those >65 years. The relationship between flavored e-cigarette use and frequency was modest (varying from 66% to 73% in PATH and 60% to 76% in NTBM, across a range of use frequencies), and was less consistent between surveys. E-cigarette users who were former smokers were more likely to use flavors (70%/74%), compared to current smokers (66%/66%); among current smokers, those who smoked less frequently were more likely to use flavors (80%/82% of <weekly smokers, 63%/58% of daily smokers). These patterns suggest adoption of flavored e-cigarettes increases with decreasing smoking (including cessation). The relatively close agreement between the two surveys - despite differences in definitions of use, sampling, and data collection methods - suggests these findings are robust, and that similar estimates can be obtained via different methods.

Note that the 71st TSRC had to be rescheduled due to a hurricane. As a result, several papers were withdrawn, and several had to be presented by the author's colleagu

112. VARIATIONS IN INTENSITY OF E-CIGARETTE USE, SMOKING HISTORY, AND DEMOGRAPHICS AMONG PAST-30-DAY E-CIGARETTE USERS. Saul Shiffman[1], Mark Sembower[1], Mimi KIM[2] and Geoffrey Curtin[2]; [1]Pinney Associates, Pittsburgh, PA, USA; [2]RAI Services Company, Winston-Salem, NC, USA

Analyses examining e-cigarette use often define "users" as those with any past-30-day use, ignoring substantial variations in use patterns and leading to incorrect perceptions of users. We used data from a national sample of 153,019 U.S. adults (16,987 e-cigarette users), surveyed from online research panels during 2013-2016, to define and describe levels of e-cigarette use based on frequency and amount. Among adults reporting any past-30-day e-cigarette use, 10% used on only one day/month, and 28% used less than weekly (≤4 days/month); conversely, 22% used daily/nearly daily (≥27 days/month). Amount of use also varied, with median use being one use/day; >75% of users reported ≤5 uses/day. Individual characteristics varied substantially among categories of past-30-day users. Daily users were older than non-daily users (43 versus 36 years; this and all comparisons cited, p<0.005). Young adults (18-24 years) were more likely to report having used e-cigarettes during the past 30 days (OR=1.8), but more likely to have used e-cigarettes on one day/month (OR=2.2) and to report low levels of use (<5 uses/day) (OR=1.6). Older adults (≥45 years) were more likely to use daily (OR=2.2), and to report ≥10 uses/day (OR=1.8). Although women were less likely to report past-30-day use (OR=0.7), they were more likely to use daily (OR=1.4). Overall, >89% of past-30-day e-cigarette users had been established smokers (100+ cigarettes), and those who used e-cigarettes more frequently were significantly more likely to have quit smoking; 56% of those reporting ≥10 uses/day were no longer smoking. These data document large variations in frequency and amount of e-cigarette use among past-30-day-users, suggesting that a more discriminating and detailed characterization of users is necessary to understand e-cigarette use.

113. COMPARISION OF BACKGROUND RISKS WITH RISKS ESTIMATED FROM CONSTITUENTS IN TOBACCO PRODUCTS. Charlene LIU and Kristin Marano, RAI Services Company, Winston-Salem, NC, USA

The U.S. Food and Drug Administration (FDA) has identified a list of harmful and potentially harmful constituents (HPHCs) in tobacco products and tobacco smoke for reporting under the Family Smoking Prevention and Tobacco Control Act. The HPHCs include some chemicals approved by FDA as food flavorings (acetaldehyde, o-, m- and p-cresols) or ingredients used in food packaging (acrylonitrile), and active ingredients or excipients used in drugs and consumer products (formaldehyde and polycyclic aromatic hydrocarbons [PAHs]). Some HPHCs are found in the soil, water, and air as a ubiquitous presence in the human environment (arsenic, acrolein, cadmium, benzene, formaldehyde, PAHs, toluene). The objective of this study is to compare potential excess lifetime cancer risk (ELCR) associated with HPHCs from cigarette smoke and smokeless tobacco products with ELCR associated with exposure to the same chemicals from background ambient air, drinking water, and food sources. U.S. market survey data of HPHCs in machine-generated cigarette smoke and smokeless tobacco products, and chemical concentrations of HPHCs in the environment and food are obtained from the literature. Potential cancer risks are evaluated following standard quantitative risk assessment process, utilizing toxicity values from the U.S. Environmental Protection Agency (EPA) recommended hierarchy sources and standard default exposure factors from EPA and FDA. These were supplemented as necessary with assumptions to develop estimates of tobacco use and accompanying exposures. Results of the study indicate that estimated ELCR associated with HPHC from tobacco products are similar to, or in some cases less than, cancer risks attributed to the same chemicals from air, food, and drinking water sources. While no tobacco product is safe or without risks, these findings indicate that potential cancer risks from exposure to many HPHCs in tobacco products, as they are estimated by contemporary procedures, are minimal and do not increase cumulative ELCR in a meaningful way, when associated with background intake of the same chemical.

116. QUANTIFYING TOBACCO LEAF COMPOUNDS BY VISIBLE LIGHT ANALYSIS. Makoto YOSHIZAKI; Japan Tobacco, Yokohama, Japan

The quality of tobacco leaf is related to its constituents. However, a simple and convenient method to characterize leaf compounds during cultivation, harvest, and curing has not been developed. Visible light analysis is a simple and fast method to measure leaf color, which is an important factor in the grading of

Note that the 71st TSRC had to be rescheduled due to a hurricane. As a result, several papers were withdrawn, and several had to be presented by the author's colleague.

tobacco leaves. Previous research has shown that tobacco leaf color is mainly derived from carotenoids, flavonoids, and chlorophylls. Carotenoids are degraded during cultivation and curing processes, yielding the compounds responsible for tobacco's aroma. In this study, we investigated degraded carotenoids using visible light analysis. Thirty-one flue-cured Virginia tobacco samples with different leaf positions from 13 countries were pulverized, and the color of the tobacco leaves was measured by spectrophotometry. The colors were represented by the L*a*b* color space (L*: Lightness; a*: Green-red chromaticity; b*: blue-yellow chromaticity). Twelve degraded carotenoids were selected as major compounds and analyzed by gas chromatography/mass spectrometry. Multiple regression analysis was performed to estimate the abundance of the selected compounds from color parameters, and the statistical significance of estimated partial regression coefficients was tested. Visible light analysis was effective in measuring five compounds. The L* degree and a* degree effectively estimated the levels of 3-hydroxy-7,8-dihydro-beta-ionol and blumenol C. However, the highest coefficient of determination was only 53%, implying that other pigments can interfere with visible light analysis.

117. INVESTIGATION ON THE ROLE OF COPPER (II) ON NITROSAMINE FORMATION FROM NICOTINE. Syed HAQUE and Socrates Jose P. Canete; University of Kentucky, Lexington, KY, USA

The role of Cu2+ in nicotine nitrosation reaction was investigated using High-Performance Liquid Chromatography (HPLC), Nuclear Magnetic Resonance (NMR) and UV-vis spectroscopy. The nitrosation reaction of nicotine was carried out in both "clean" citrate phosphate buffer and in tobacco matrix. Titration with Cu2+ monitored by NMR and pH dependent UV-vis profiles strongly indicate that Cu2+ binds irreversibly to the pyridine nitrogen of nicotine. Studies on nitrosation kinetics reveal that in presence of Cu2+, the activation energy of the nitrosation reaction was increased suggesting unfavorable nitrosamine-formation kinetics. NMR and kinetics data strongly suggest that by binding irreversibly with the pyridine moiety of nicotine, Cu2+ imparts partial inhibition of nitrosamine formation from nicotine. This presentation will demonstrate the extent of nicotine nitrosation and role of copper on in the nitrosation reaction.

118. PON AND NNK CHANGES IN CURED BURLEY TOBACCO WITH STORAGE. Ying WU, Huihua Ji, Franklin Fannin and Lowell Bush; University of Kentucky, Lexington, Kentucky, USA

4-(N-methyl-N-nitrosamino)-1-(3-pyridyl)-1-butanone (NNK) is considered to be the most potent carcinogen of the TSNAs that are a class of nitrosamines that occur in tobacco products. Previous kinetics studies have shown the generation of NNK is pH dependent and temperature can affect the reaction rate from pseudo-oxynicotine (PON) to NNK. A very rapid formation of NNK from PON was measured using 200 μmole ml-1 nitrite in pH 3 buffer at 37°C. Nitrate and nitrite enhancement for the increase in TSNAs during warm temperature storage and moisture content is not well understood. The objectives of this study were to determine the PON and NNK changes during the storage time at four different moisture conditions. Cured ground burley tobacco which contained 7, 12, 18 and 28 % moisture were stored with excessive nitrate and nitrite in separated sealed containers at 60 °C. Nitrate level did not effect NNK accumulation in 15 days. Excessive nitrite increased NNK levels from 5.5-35X compared to controls in 15 days under different moisture levels. The greatest increase of NNK occurred with 12% moisture (~35X). PON levels were increased 4-6-fold with all moisture levels during 15 days of treatment. Measured PON increased rapidly during the first two days, especially at 7% moisture. The results indicate that NNK accumulation was altered by temperature and moisture content of tobacco.

123. THE REGULATION OF MICROORGANISM ON THE FORMATION OF TSNAS IN TOBACCO. ZOU Congming, Xia Zhenyuan, Lei Liping, Huang Wei, Zhao Gaokun, Jin Yan and Wu Qun; Yunnan Academy of Tobacco Agricultural Sciences, Kunming, China

TSNAs in tobacco are directly approved to be the carcinogens which is getting more and more attention than before. The regulation of TSNAs is becoming one of the most important topics of tobacco harm reduction. It has been shown that TSNAs formation is significantly interacted with the microbes in the flue-curing process. Therefore, the objective of this study is to explore the effects of microbial communities on the formation of tobacco TSNAs and further confirm the relationship between them. Three series of studies

Note that the 71st TSRC had to be rescheduled due to a hurricane. As a result, several papers were withdrawn, and several had to be presented by the author's colleagu

were conducted to explore the effects of microbial communities on the TSNAs formation, including fresh leave cultivation, air-curing process, cured leave cultivation. The sequencing method was used to identify the type and function of each strain. This study found that a microorganism (J54) could significantly increase the content of TSNAs in tobacco, while the other (05-5402) could significantly reduce the content of TSNAs to a large extent and other microorganisms were less obvious. Sequencing results show that J54 belongs to Pseudomonas sp., and 05-5402 belongs to bacillus pumilus. This study clarifies the positive and negative effects of several strains on tobacco TSNAs formation and would play an important role in reducing carcinogens for tobacco. The reduction effect can be directly used in the development of biological method to reduce harmful contents, at the same time, the microorganism (J54) could provide the target for reducing harm study in tobacco.

124. ARE AVAILABLE TEST METHODS FOR THE DETERMINATION OF AMMONIA IN MAINSTREAM CIGARETTE SMOKE FIT FOR THE ANALYSIS OF CIGARS? Yevgeniya PREPELITSKAYA, Kathleen Spangler, Jennifer H. Smith, Karen Avery, James Wilkinson, Matt Melvin and John H. Miller IV; Altria Client Services, Richmond, VA USA

In May 2016, the U.S. Food and Drug Administration (FDA) issued a final rule to deem cigars to be subject to the Federal Food, Drug, and Cosmetic Act, as amended by the Family Smoking Prevention and Tobacco Control Act. As part of this regulation, the FDA will require manufacturers to report the quantities of Harmful and Potentially Harmful Constituents (HPHCs) in cigar filler and smoke. Standardized methods do exist for the analysis of ammonia in cigarette smoke; however, these methods may not be fit for purpose for the analysis of cigars. Cigars vary widely in blend composition and size compared to cigarettes, which could further complicate the analysis of these products. CORESTA Recommended Method No. 83 Determination of Ammonia in Mainstream Cigarette Smoke by Ion Chromatography (CRM 83) is a standardized method used for ammonia in cigarette mainstream smoke. The results of this work indicate that the CORESTA Recommended Method may not be appropriate for all cigar products. In our evaluation of the trapping efficiency, we observed higher yields of ammonia in mainstream smoke with increasing concentration of acid in the trapping solution. In addition, the amount of ammonia detected in sample extracts increased over time faster with higher concentrations of acid. These observations suggest that under acidic conditions, another component in mainstream smoke may break down into ammonia. We also observed differences in these effects depending on the cigar blend composition. These results indicate that specialized methods need to be developed for the analysis of ammonia in cigar smoke.

125. DETERMINATION OF HARMFUL AND POTENTIALLY HARMFUL CONTAMINANTS IN TOBACCO SMOKE USING STANDARD REFERENCE MATERIAL 3222. Walter B. WILSON and Lane C. Sander, National Institute of Standards and Technology, Gaithersburg, MD, USA

Under the Food, Drug, and Cosmetic Act (FD&C) all tobacco manufacturers and importers are required to report the levels of harmful and potentially harmful constituents (HPHCs) found in their tobacco products and tobacco smoke. Studies have reported the detection of up to 7000 different chemical components that can be inhaled through tobacco smoke such as nicotine, tar, nitrosamines, and polycyclic aromatic hydrocarbons. In 2012, the Food and Drug Administration (FDA) included in the Federal Register a list of 93 HPCHs suspected to be found in tobacco products and tobacco smoke. To support the Family Smoking Prevention and Tobacco Control Act, the National Institute of Standards and Technology has recently collaborated with the Center for Tobacco Products at the FDA to develop a new low nicotine tobacco filler Standard Reference Material (SRM 3222) to support the analysis of tobacco products. This new tobacco material has certified and reference mass fraction values for nicotine, N-nitrosonornicotine (NNN), 4-(methylnitrosamino)-1-(3-pyridyl-1-butanone (NNK), and moisture. In the current study, approximately 10 g of SRM 3222 was burned to produce tobacco smoke. Several different smoke collection apparatuses and organic solvents were evaluated to allow for the highest amount of tobacco smoke to be collected. After filtration, the homogenous solution was screened via gas chromatography/mass spectrometry for the identification of many HPHCs in the tobacco smoke.

Note that the 71st TSRC had to be rescheduled due to a hurricane. As a result, several papers were withdrawn, and several had to be presented by the author's colleague.

127. FUMEX: LIGHT SCATTERING SENSOR FOR THE ANALYSIS OF ELECTRONIC CIGARETTE AEROSOLS. Weiling LI[1], Qiang Wang[2]; Peter Lipowicz[1], Wilhelm. Dunkhorst[3] and Wolfgang Koch[3]; [1]Altria Client Services, Richmond, VA USA; [2]Lancaster Laboratories, c/o Altria Client Services, Richmond, VA USA; [3]Fraunhofer ITEM, Hannover, Germany

A new sensor based on light scattering has been designed to perform highly time resolved measurements of the mass concentration and the mass median diameter (MMD) of aerosols formed in e-cigarettes. Aerosols generated in e-cigarettes are spherical, submicron particles with well-defined constant optical properties. The Fumex sensor measures scattered light from the e-cigarette aerosol at two polarizations. The mass concentration and mass median diameter of the aerosol is calculated from the light scattering signals assuming a fixed geometric standard deviation and refractive index equal to that of the bulk e-liquid. The light source used in Fumex is a laser diode emitting polarized light at 680 nm. The scattered light is detected at 90° using a semi conductor photodetector. The validation of this sensor was carried out with an impactor for MMD and with filter gravimetric measurement for mass concentration. Good correlation for both parameters was observed, R-squared value 0.97 for MMD and 0.95 for mass concentration. Application ranges for Fumex are: mass concentration range of 1-50 g/m3, MMD of 0.2 -1.5 µm, 100ms time resolution, and 0.2-3 l/min flow rate. This sensor has been used to measure particle size of e-cigarettes with different carriers, and a variety of commercial e-cigarettes. Results were compared to measurement done with Spraytec and impactors. Measurement with different commercial e-cigarettes using this sensor will also be discussed in the presentation. Fumex can provide fast measurement of mass concentration and real-time monitoring of aerosol generation during a single puff, which allows better understanding of aerosol formation in the e-cigarette.

129. A COMPUTATIONAL MODEL TO CHARACTERIZE THE VITROCELL® CELL EXPOSURE SYSTEM FOR EVALUATION OF AEROSOLS. Nicolas CASTRO[1], Ali A. Rostami[1], Michael J. Oldham[1], Yezdi B. Pithawalla[1]; Francesco Lucci[2] and Arkadiusz Kuczaj[2]; [1]Altria Client Services, Richmond, VA, USA; [2]Philip Morris International, Neuchâtel, Switzerland

In vitro exposure systems can be used as tools for toxicological assessment of e-cigarette aerosols. The VITROCELL® exposure system is designed such that exposure of cell cultures to the aerosol occurs at the air/liquid interface, which is relevant to e-cigarette use. It is difficult to experimentally quantify the actual cell dose applied in the VITROCELL® system, as it depends on multiple parameters such as system geometry, particle size and distribution, air flow-rate, exposure level and duration, etc. A computational fluid dynamics aerosol tracking and deposition model that employs Lagrangian particle tracking method has been developed to quantify deposition rates of particles on the air/liquid interface in a VITROCELL® 24/48 system. The system consists of a 6 mm diameter main line carrying the aerosol, with six smaller 3 mm diameter tubes (trumpets) branching down to the cell exposure plates. Results of simulations for a main line air flow rate of 1 L/min and trumpet flow rates of 1- 4 mL/min will be discussed. Simulations were performed for inert non-reacting solid particles with a range of diameters between 0.5 and 4.5 µm and densities of 1050kg/m3. The impact of particle size and air flow rate on deposition efficiency on the cell plates was explored. Results show that for a trumpet to main air flow ratio of 2/1000, the fraction of inlet particles deposited on the cell exposure plates is less than 0.001. This can be attributed to (1) lower particle concentration near the wall in the main line and (2) carryover of particles by the trumpet air out of the system. Once validated, the model will be used to quantify cell exposure from different e-cigarette aerosol streams.

Note that the 71st TSRC had to be rescheduled due to a hurricane. As a result, several papers were withdrawn, and several had to be presented by the author's colleagu